Transport in Shale Reservoirs

Transport in Shale Reservoirs

KUN SANG LEE
Hanyang University
Seoul, South Korea

TAE HONG KIM
Hanyang University
Seoul, South Korea

Gulf Professional Publishing
An imprint of Elsevier

TRANSPORT IN SHALE RESERVOIRS   ISBN: 978-0-12-817860-7
Publisher: Brian Romer
Senior Acquisition Editor: Katie Hammon
Editorial Project Manager: Ali Afzal-Khan
Project Manager: Poulouse Joseph
Cover Designer: Miles Hitchen

50 Hampshire Street, 5th Floor, Cambridge, MA 02139, United States

The Boulevard, Langford Lane, Kidlington, Oxford, OX5 1GB, United Kingdom

Preface

As global energy consumption is steadily growing, the hydrocarbon sources from unconventional shale reservoirs are increasing rapidly. Although the shale industry has made rapid progress in recent years, there is still a lack of knowledge both in industrial and academic fields. Because shale reservoirs have distinctive features different from those of conventional reservoirs, an accurate evaluation on the behavior of shale reservoirs needs an integrated understanding on the characteristics and transport of reservoir and fluids. This book aims to present a comprehensive and mathematical treatment of characterization and modeling of shale reservoirs.

While several books exist for shale gas reservoir, many of them focused on geological, economic, environmental aspects. There has been a void for a comprehensive book that focuses on the transport phenomena through shale reservoirs. This book emphasizes all relevant aspects of petrophysical characteristics and their impact on transport mechanisms. The book also discusses a systematic approach in the modeling of shale reservoirs based on the complicated transport mechanisms. The authors' desire is that the information in this book provides a clear presentation of the transport principles in shale reservoirs, state-of-the-art technology on the modeling, applications in the real field, and ideas for future research.

Chapter 1 serves as an introduction to the topic by tracing the development and current status of shale reservoirs. Chapter 2 reviews petrophysical characteristics of shale reservoirs such as lithology, mineral composition, organic matter, pore geometry, and fracture-matrix system. Chapter 3 discusses major transport mechanisms in shale reservoirs including non-Darcy flow, gas sorption, molecular diffusion, geomechanics, and phase behavior in nanopores. Chapter 4 focuses on the simulation of shale gas and oil reservoirs. Chapter 5 presents emerging technologies in shale reservoirs including multicomponent transport in the CO_2 injection process and consideration of organic material in shale reservoir simulation.

This book would never have been published without the able assistance of Elsevier staff for their patience and their excellent editing job. We shall appreciate any comments and suggestions.

Kun Sang Lee
Seoul, Korea

Contents

CHAPTER 1

Introduction

ABSTRACT

As conventional hydrocarbon resources have depleted and global energy consumption has grown steadily, interest of unconventional resources has increased steadily. Especially, production of shale gas and oil resources increases rapidly after development of hydraulic fracturing and horizontal drilling techniques in North America. Even though shale industry has grown dramatically, fluid flow in shale reservoirs still has not been fully understood. Unconventional shale reservoirs showed complex flow mechanisms compared with conventional reservoirs. Shale reservoirs have intricate petrophysical characteristics such as various lithologies and mineral composition, organic component, tiny pore geometry, and natural fracture system, and they affect fluid behavior significantly. Non-Darcy flow, adsorption/desorption, fluid flow and phase behavior change in nanoscale pores, molecular diffusion, and stress-dependent deformation should be considered for evaluation of fluid flow in shale reservoirs. For accurate understanding of transport in shale reservoirs, this book presents comprehensive study of petrophysical characteristics, transport mechanisms, and application in field-scale reservoir simulation.

Because of rapid depletion of conventional resources and increase of global energy consumption, conventional hydrocarbon resources cannot satisfy the energy demand. According to World Energy Outlook 2017 (IEA, 2017), global energy will expand by 30% between today and 2040. Although various researches have been performed to develop sustainable and renewable energy, it is still significantly expensive to be commercialized. Consequently, unconventional oil and gas resources have attracted considerable attention in recent decades. Unconventional oil and gas are hydrocarbon resources that cannot be produced with conventional extraction techniques. Fig. 1.1 shows the hydrocarbon resources pyramid. Unconventional resources include tight oil, shale oil, oil shale, coalbed methane (CBM), tight gas, shale gas, and gas hydrates. Oil shale is a rock that includes significant amounts of organic matters such as

kerogen. CBM is natural gas contained in the coal seam. Gas hydrate is an icelike form of crystalline water-based solid that contains gas molecules in its molecular cavities. In general, unconventional resources are more expensive and more difficult to extract and process. As shown in Fig. 1.1, resources in the lower part of the pyramid present a higher amount of reserves, the increased difficulty of production, and higher development cost than resources in the upper part. Among various unconventional resources, shale gas and oil are the most commercialized resources so far.

GEOLOGIC FEATURES OF SHALE RESERVOIRS

Recently, shale reservoirs have received significant attention because of their potential to supply the world with an immense amount of energy and the depletion of conventional reservoirs. Because shale reservoirs have various specific features different from conventional reservoirs, there is still a lack of understanding for them. Conventional reservoirs are comprised of source rock, reservoir rock, trap, and seal. In a typical source rock, some of the hydrocarbons are driven out and migrate into reservoir rock under the traps. In shale oil and gas reservoirs, the generated hydrocarbon cannot be migrated, and the source rock itself becomes the reservoir rock because of its tight features. Shale is a fissile and laminated sedimentary rock mainly composed of clay-sized mineral grains. Normally, in a broad sense, shale reservoir rock contains clastics (quartz, feldspars, and micas), carbonates (calcite, dolomite, and siderite), clay minerals (montmorillonite, illite, smectite, and kaolinite), pyrite, and the other minor minerals (Passey, Bohacs, Esch, Klimentidis, & Sinha, 2010; Quirein et al., 2010; Ramirez, Klein, Ron, & Howard, 2011; Sondergeld, Newsham, Comisky, Rice, & Rai, 2010). Especially, black-colored shale rock includes organic matters such as kerogen and is an essential source of shale oil and gas.

Amount of organic materials, which indicates the capacity to produce and store hydrocarbons, is significant in shale reservoirs. As a conventional source rock,

Transport in Shale Reservoirs. https://doi.org/10.1016/B978-0-12-817860-7.00001-2

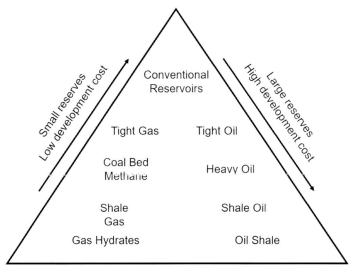

FIG. 1.1 Hydrocarbon resources pyramid.

roughly 0.5% total organic carbon (TOC) is considered as a minimum or threshold. For shale reservoirs, approximately 2% is regarded as a minimum TOC for commercial production, and it may exceed 10%–12% in some reservoirs. Kerogen types are primarily classified into four categories (Tissot & Welte, 1984). Based on the character, elemental contents, and depositional environments of kerogens, kerogens are classified to types I, II, III, and IV. Investigation of these kerogen types is important to understand the processes of storage, retention, and release of hydrocarbons. Commonly, oil is produced from shale reservoir containing kerogen types I and II, and gas is produced from a reservoir containing kerogen type III. Thermal maturity, which indicates maximum temperature exposure of rock and extent of temperature-time-driven reactions, is another critical parameter in shale reservoirs. The organic materials whose vitrinite reflectance is lower than $0.65\%R_o$ are considered as immature organic matters (Mani, Patil, & Dayal, 2015). Thermally mature organic matters, which present $0.6\%–1.35\%R_o$ of vitrinite reflectance, commonly produce oil. The postmature organic materials, which show higher than $1.5\%R_o$ of vitrinite reflectance, generate wet and dry gas (Mani et al., 2015; Tissot & Welte, 1984).

Unconventional shale reservoirs also show complex pore geometry because of various lithology, mineral composition, and organic matters. Pore networks of shale formations consist of organic matter pores, inorganic material pores, and natural fracture system (Fig. 1.2). Organic matter pores can be divided into primary organic pores and secondary organic pores. Secondary organic matter pores also can be divided into organic matter bubble and spongy pores. Inorganic pores can be divided into interparticle and intraparticle mineral pores. The pore size of shale reservoirs ranges from nanometers to micrometers. The complexity of pore geometry and fracture networks significantly affects the behavior of hydrocarbon in shale reservoirs. Understanding of geologic features of shale rock is an important prerequisite for analysis of transport in shale reservoirs.

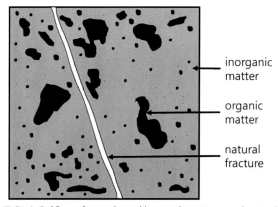

FIG. 1.2 View of organic and inorganic matters and natural fracture in shale reservoirs.

SHALE BOOM

Although the commercialization of shale oil and gas was performed in the 2000s, the existence of shale reservoirs was discovered in about two centuries ago, and there are tremendous exertions to develop the shale reservoirs. In 1821, the shallow shale gas reservoir was discovered and drilled in the Devonian Dunkirk Shale in Chautauqua County, New York (Wang, Chen, Jha, & Rogers, 2014). After this discovery, numerous shallow shale wells were drilled along the Lake Erie shoreline (Hill, Lombardi, & Martin, 2004). In 1863, shale gas reservoirs were discovered in the western Kentucky part of Illinois Basin. By 1920s, drilling of shale gas was established to West Virginia, Kentucky, and Indiana (Wang et al., 2014). In the 1940s, hydraulic fracturing is first used to stimulate gas wells operated by Pan American Petroleum Corporation in Grant County, Kansas.

After the oil crisis in the 1970s, the US federal government has invested in research and development of shale gas for alternative energy of oil. In late 1976, the US Department of Energy performed Eastern Gas Shale Project, which lasted to 1992. In this project, a series of geologic, geochemical, and petroleum engineering studies were conducted to evaluate the gas potential and to enhance gas production from extensive Devonian and Mississippian organic-rich black shale within the Appalachian, Illinois, and Michigan basins of the eastern United States (NETL, 2011). In the meantime, private oil companies also invested in unconventional natural gas due to the high oil prices (Cleveland, 2005; Henriques & Sadorsky, 2008). However, deep shale reservoirs such as Barnett Shale in Texas and Marcellus Shale in Pennsylvania were not considered as economically feasible reservoirs owing to ultralow permeability at that time.

For the economic production of shale gas, several pioneering oil enterprises had tried to perform hydraulic fracturing. From the 1980 to 1990s, the Mitchell Energy & Development Corporation tested various processes of hydraulic fracturing to produce natural gas in Barnett shale, eventually finding the relevant technique economically. The hydraulic fracturing technique developed by Mitchell Energy & Development Corporation has been widely used by the oil company, and it changed the face of the petroleum industry in the 2000s. In other words, exertions of US government and various companies for several decades have made tremendous shale boom in these days.

Annual Energy Outlook 2018 expected that production of natural gas and oil in United States will increase due to the result of continued development of shale gas and tight oil plays (EIA, 2018). EIA expected tight oil

and shale gas account for 65% and 75% of crude oil and natural gas production, respectively, in the United States as shown in Figs. 1.3 and 1.4. In addition, shale resources are abundant in the world. Fig. 1.5

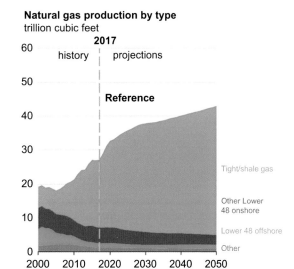

FIG. 1.3 Expectation of natural gas production in the United States by 2050. (Source: U.S. Energy Information Administration (Feb 2018).)

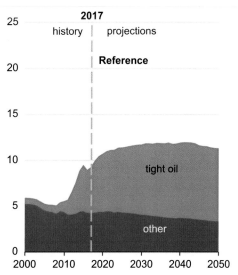

FIG. 1.4 Expectation of crude oil production in the United States by 2050. (Source: U.S. Energy Information Administration (Feb 2018).)

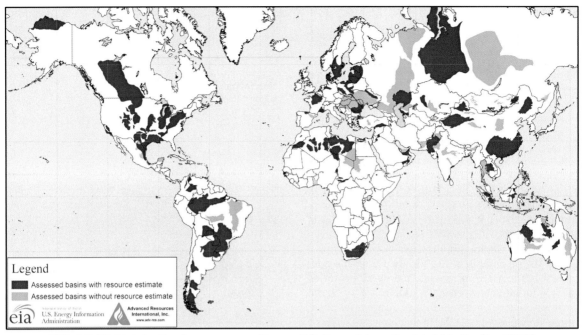

FIG. 1.5 World shale oil and gas resources. (Source: U.S. Energy Information Administration (Jun 2013).)

presents a map of basins assessed shale oil and gas formations (EIA, 2013). According to EIA (2015), shale resources are located in 46 countries. EIA report also estimates unproved technically recoverable tight oil of 419 billion barrels and shale gas of 7577 trillion cubic feet in the world (EIA, 2015). Table 1.1 presents the estimates of unproved technically recoverable shale gas and tight oil categorized by continents (EIA, 2015). As shown in Fig. 1.5 and Table 1.1, the potential of shale oil and gas is tremendous due to an abundant amount of reserves in a wide region. Although commercialization of shale resources is still limited in the United States in these days, estimated reserves are higher in other continents. Therefore, many countries are interested in these cheaper and cleaner energy sources from shale reservoirs.

TABLE 1.1
Estimates of Unproved Technically Recoverable Shale Gas and Tight Oil in the World

Region	Gas (Tcf)	Oil (Billion bbl)
North America	1741	100
South America	1433	60
Europe	907	93
Asia/Australia	1802	72
Africa	1406	54
Middle East	288	40

Reproduced from EIA (2015). *World shale resource Assessments.* https://www.eia.gov/analysis/studies/worldshalegas/.

TRANSPORT MECHANISMS IN SHALE RESERVOIRS

Unconventional shale reservoirs have significantly different characteristics compared with conventional reservoirs. Therefore, flow behavior in shale reservoirs cannot be explained with conventional processes. The most critical feature of shale reservoir is tight reservoir condition. Shale reservoirs have low permeability and low porosity matrix condition. Generally, shale reservoirs have nano-Darcy to micro-Darcy scale of permeability and less than 10% of porosity. To extract oil and gas from these tight formations, the hydraulic fracturing should be performed. Millions of gallons of water with proppant and chemicals are injected to break the shale formation. Because injected fluid induces fractures and rejuvenates the existing natural fractures around the wellbore, oil and gas are extracted through the high permeability fracture networks as shown in

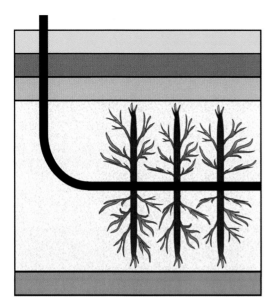

FIG. 1.6 View of hydraulically fractured horizontal well.

Fig. 1.6. In addition, because of the relatively thin and widespread reservoir structure of shale reservoirs, horizontal wells are used. Horizontal well increases the contact area of the wellbore that improves fluid flow from the reservoir to the wellbore and decreases completion cost.

Even though most conventional reservoir system can be computed by Darcy's law (1856), fluid behavior in shale reservoir system cannot be directly calculated. In hydraulic fractures, where the fluid velocity is very high, inertial forces cannot be ignored compared with viscous forces. In this situation, the pressure behavior deviates from Darcy's law so that Forchheimer equation (1901) with non-Darcy coefficient is used to calculate pressure responses. In shale gas reservoirs, the hydrocarbon gas is stored in free and adsorbed states. Some gas exists in pore spaces of the matrix and fractures, and the other gas is adsorbed on the surface of organic matters. Because of nano- to microscale pore size of shale matrix, slippage effect and Knudsen diffusion were presented (Javadpour, 2009; Javadpour, Fisher, & Unsworth, 2007). In addition, small pore size induces high capillary pressure and changes the behavior of fluid properties. This effect is called capillary condensation or confinement effect. Molecular diffusion should also be considered because of the ultratight condition of reservoirs. In addition, the conductivity of fractures is susceptive to change of stress and strain caused by pressure variation. Reservoir deformation in shale reservoir significantly affects the productivity of the fractured well.

OBJECTIVES

In most field cases of shale reservoirs, operators commonly cannot consider specific fluid behavior in the subsurface system. They concentrate on hydraulic fracturing and horizontal drilling. However, for the accurate prediction of oil and gas production in shale reservoirs, exact transport mechanisms, which are significantly different from a conventional reservoir, should be thoroughly understood. Transport mechanisms occurred in shale reservoir include non-Darcy flow, adsorption/desorption, microscale flow, molecular diffusion, stress-dependent deformation, and confinement effects. To help readers understand principles of these mechanisms, the scope of this book ranges from the basic geology of shale reservoir to mathematical formulation of various transport mechanisms. Because studies for shale reservoir are still in an early stage and may not reach a consensus, this book tried to present all information from established basic theories to various up-to-date theories. In addition, based on multiple transport mechanisms, numerical simulations were performed in several field examples of shale gas and oil reservoirs. Finally, for the more advanced subject, CO_2 injection in shale reservoirs and effects of organic matters are presented.

REFERENCES

Cleveland, C. J. (2005). Net energy from the extraction of oil and gas in the United States. *Energy, 30*(5), 769–782. https://doi.org/10.1016/j.energy.2004.05.023.

Darcy, H. (1856). *Les fontaines publiques de la ville de Dijon: Exposition et application des principes a suivre et des formules a employer dans les questions de distribution d'eau; ouvrage terminé par un appendice relatif aux fournitures d'eau de plusieurs villes au filtrage des eaux et a la fabrication des tuyaux de fonte, de plomb, de tole et de bitume: Victor Dalmont, Libraire des Corps imperiaux des ponts et chaussées et des mines.*

EIA. (2013). *Technically recoverable shale oil and shale gas resources: An assessment of 137 shale formations in 41 countries outside the United States.* Washington, DC: U.S. Department of Energy.

EIA. (2015). *World shale resource assessments.* https://www.eia.gov/analysis/studies/worldshalegas/.

EIA. (2018). *Annual energy outlook 2018.* Washington, DC: U.S. Department of Energy.

Forchheimer, P. (1901). Water movement through the ground. *Zeitschrift Des Vereines Deutscher Ingenieure, 45,* 1781–1788.

Henriques, I., & Sadorsky, P. (2008). Oil prices and the stock prices of alternative energy companies. *Energy Economics, 30*(3), 998–1010. https://doi.org/10.1016/j.eneco.2007.11.001.

Hill, D. G., Lombardi, T. E., & Martin, J. P. (2004). Fractured shale gas potential in New York. *Northeastern Geology and Environmental Sciences, 26,* 22.

IEA. (2017). *World energy outlook 2017.*

Javadpour, F. (2009). Nanopores and apparent permeability of gas flow in mudrocks (shales and siltstone). *Journal of Canadian Petroleum Technology, 48*(08), 16–21. https://doi.org/10.2118/09-08-16-DA.

Javadpour, F., Fisher, D., & Unsworth, M. (2007). Nanoscale gas flow in shale gas sediments. *Journal of Canadian Petroleum Technology, 46*(10), 55–61. https://doi.org/10.2118/07-10-06.

Mani, D., Patil, D. J., & Dayal, A. M. (2015). Organic properties and hydrocarbon generation potential of shales from few sedimentary basins of India. In S. Mukherjee (Ed.), *Petroleum geosciences: Indian contexts* (pp. 99–126). Cham: Springer International Publishing.

NETL. (2011). *Shale gas: Applying technology to solve America's energy challenges*. Washington, DC: U.S. Department of Energy.

Passey, Q. R., Bohacs, K., Esch, W. L., Klimentidis, R., & Sinha, S. (2010). From oil-prone source rock to gas-producing shale reservoir — geologic and petrophysical characterization of unconventional shale gas reservoirs. In *Paper presented at the international oil and gas conference and exhibition in China, Beijing, China, 2010/1/1*. https://doi.org/10.2118/131350-MS.

Quirein, J., Witkowsky, J., Truax, J. A., Galford, J. E., Spain, D. R., & Odumosu, T. (2010). Integrating core data and wireline geochemical data for formation evaluation and characterization of shale-gas reservoirs. In *Paper presented at the SPE annual technical conference and exhibition, Florence, Italy, 2010/1/1*. https://doi.org/10.2118/134559-MS.

Ramirez, T. R., Klein, J. D., Ron, B., & Howard, J. J. (2011). Comparative study of formation evaluation methods for unconventional shale gas reservoirs: Application to the Haynesville shale (Texas). In *Paper presented at the North American unconventional gas conference and exhibition, The Woodlands, Texas, USA, 2011/1/1*. https://doi.org/10.2118/144062-MS.

Sondergeld, C. H., Newsham, K. E., Comisky, J. T., Rice, M. C., & Rai, C. S. (2010). Petrophysical considerations in evaluating and producing shale gas resources. In *Paper presented at the SPE unconventional gas conference, Pittsburgh, Pennsylvania, USA, 2010/1/1*. https://doi.org/10.2118/131768-MS.

Tissot, B. P., & Welte, D. H. (1984). *Petroleum formation and occurrence*. Springer Berlin Heidelberg.

Wang, Q., Chen, X., Jha, A. N., & Rogers, H. (2014). Natural gas from shale formation — the evolution, evidences and challenges of shale gas revolution in United States. *Renewable and Sustainable Energy Reviews, 30*, 1–28. https://doi.org/10.1016/j.rser.2013.08.065.

CHAPTER 2

Petrophysical Characteristics of Shale Reservoirs

ABSTRACT

Shale reservoirs have distinctive petrophysical characteristics compared with conventional reservoirs. Shale is a fine-grained sedimentary rock generated from the compaction of silt and clay-size mineral particles. However, commonly, shale resources represent oil and natural gas hosted in a broad spectrum of lithology from mudstone to fine-grained sandstone as well as shale. Shale reservoirs contain varying amounts of organic matters, clay mineral, quartz, carbonates, and numerous minor minerals depending on plays. Organic matters such as kerogen and bitumen play an essential role in hydrocarbon generation and gas storage. Gas can be adsorbed on the surface of organic pores and dissolved in organic matters. The pore size of shale reservoirs ranges from nanometers to micrometers. In addition, the natural fracture system of shale reservoirs affects fluid flow in shale reservoirs, especially with hydraulic fractures. For an understanding of fluid behavior in shale reservoir, accurate petrophysical characterization should be preceded.

LITHOLOGY AND MINERALOGY

Lithology is the composition or type of rock such as sandstone or limestone (Hyne, 1991). The lithology of a rock unit is a description of its physical characteristics such as color, texture, grain size, and composition (Allaby and Allaby, 1999; Bates, Jackson, & Institute, 1984). It can also be a summary of the gross physical character of rock as well as a detailed description of physical characteristics. Reservoir characterization of oil and gas reservoirs starts with identifying and quantifying the lithology and mineral composition. The difference in lithology and mineral composition can impact on reservoir productivity and efficiency of fracture stimulation. The physical and chemical features of the reservoir rock can affect the response of measuring tool for formation properties. Understanding reservoir lithology and mineral composition is the foundation from which all other petrophysical calculations are made. To make accurate petrophysical estimations of porosity, permeability, water saturation, and so forth, the various lithologies and mineral composition of the reservoir interval must be identified, and their implications should be understood. Accurate determination of lithology and mineral composition can be accomplished by a combination of conventional wireline logging or logging while drilling technologies, a variety of petrological and geochemical analyses core and cuttings, and wireline elemental spectroscopy logging (Ahmed & Meehan, 2016).

Basic Lithology and Mineralogy of Shale

Shale is a fine-grained sedimentary rock generated from the compaction of silt and clay-size mineral particles, which are normally called mud. Shale is different from other mudstones because it is fissile and laminated. Therefore, shale rock is made up of many thin layers, and the rock easily splits into small pieces along the laminations.

Shale is a rock comprised primarily of clay-sized grains, which are usually clay minerals such as montmorillonite, illite, smectite, and kaolinite. Shale also contains clastic mineral particles such as quartz, feldspar, and chert; carbonate minerals; sulfide minerals; iron oxide minerals; and organic particles. The deposition environment of shale often determines these other constituents in the rock, which often decide the color of the rock. Like most rocks, just a small percent of specific materials such as iron and organic matters can considerably alter the rock color. Shales deposited in oxygen-rich environments often contain tiny particles of iron oxide or iron hydroxide minerals such as hematite, goethite, or limonite. Just a small percent of these minerals distributed through the rock can generate the red, brown, yellow, and black colors of shale. The hematite can create a red shale. The limonite or goethite can produce a yellow or brown shale. These shales can be

Transport in Shale Reservoirs. https://doi.org/10.1016/B978-0-12-817000-7.00002-1

crushed to produce clay and cement, which can be used to make a variety of useful objects.

Black color in sedimentary rocks almost indicates the existence of organic matters. Just 1% or 2% organic matters can generate a black or dark gray color to the rock. In addition, this black color implies that the shale formed from sediment deposited in an oxygen-deficient environment. Any oxygen entering the environment quickly reacted with the decaying organic debris. If a large amount of oxygen was present, all organic debris would have been decayed. An oxygen-poor environment also provides the proper conditions for the formation of sulfide minerals such as pyrite, which is generally found in black shales. The presence of organic debris in black shales makes them the candidates for oil and gas generation. If the organic material is preserved and heated adequately after burial, oil and natural gas might be generated. Various shale gas and oil reservoirs in the United States, such as the Barnett Shale, Marcellus Shale, Haynesville Shale, Fayetteville Shale, Bakken Shale, and Eagle Ford Shale, are all dark gray or black shales.

Oil shale is a rock that contains substantial amounts of organic materials in the form of kerogen. Up to one-third of the rock can be solid kerogen in oil shale. Although liquid and gaseous hydrocarbons can be extracted from oil shale, the rock must be heated and treated with solvents. The extraction process of oil shale is usually inefficient than drilling rocks, which yield oil or gas directly into a well. Extracting the hydrocarbons from oil shale produces emissions and waste products that cause significant environmental concerns. Therefore, extensive oil shale deposits in the world have not been aggressively used yet. In this book, oil shale is out of scope.

As mentioned earlier, the definition of shale that best describes the reservoir is organic-rich and fine-grained rock. However, generally, the term shale is used very loosely, and it does not represent the lithology of the reservoir (Rokosh, Pawlowicz, Berhane, Anderson, & Beaton, 2008). Lithologic variations in shale gas and oil reservoirs indicate that natural gas and oil are hosted not only in shale but also in a broad spectrum of lithology and texture from mudstone to siltstone and fine-grained sandstone, any of which may be of siliceous or carbonate composition. The matrix mineral composition of these organic-rich shale reservoirs is typically heterogeneous containing varying amounts of kerogen, clays (montmorillonite, illite, smectite, and kaolinite), clastics (quartz, feldspar, and mica), carbonates (calcite, dolomite, and siderite), pyrite, and minor occurrences of additional minerals.

Generally, shale reservoirs present low to moderate clay volumes, variable quartz content, where decreasing quartz volume is commonly offset with increasing calcite volume. The formations typically contain up to 12%–15% by volume of kerogen, the source of methane gas present in the pore space. There is also adsorbed gas in association with the kerogen. Shale rocks contain small amounts of uranium and other radioactive elements that render the conventional gamma ray (GR) log essentially useless for quantitative interpretation (Ramirez, Klein, Ron, & Howard, 2011).

According to Jiang et al. (2016), different assemblages and proportions of laminated structures lead to different lithofacies. There are four kinds of a basic laminated structure such as organic matter-rich laminated structure, clastic-rich laminated structure, carbonate laminated structure, and clay laminated structure. Various mineral compositions determine the types and arrangements of shale pores. The rigidity of quartz is favorable for the preservation of intergranular pore. The instability of feldspar results in dissolution pores along the cleavage. Carbonate minerals are not only easily dissolved to form intergranular or intragranular dissolved pores but also easily recrystallized to form intergranular pores. Clay minerals are often curly shaped flakes with considerable intergranular pores. Different mineral compositions also control the brittleness of shale and further influence the later reservoir reconstruction. The higher content of brittle minerals such as quartz and calcite can result in the formation of natural fractures. During fracturing and stimulation, it is also easy to form complex-induced fractures and achieve the extension and connection of fracture networks. However, the elastic deformation of clay minerals during fracturing tends to block the channels, which is undesirable for reservoir stimulation. According to the investigation of the mineral compositions of shale reservoirs in different regions (Jiang et al., 2016), the marine shale usually has high siliceous content. The lacustrine shale usually has a high content of carbonate minerals, and the content of clay mineral is generally lower than 50%.

The multiple rock types of organic-rich shale imply that there are numerous gas storage mechanisms. Gas may be adsorbed on organic materials and stored as a free gas in micropores and macropores. Solute or solution gas, which may be held in micropores and nanopores of bitumen, may be an additional source of gas, although normally this is thought to be a minor component. Free gas may be the most dominant source of production in a shale gas reservoir. Determining the percentage of free gas, solute gas, and desorbed gas is essential for resource and reserve evaluation and is a significant issue in gas production and reserve calculations, as desorbed gas diffuses at a lower pressure than free gas. Some of the clay minerals in shale are also able to absorb or adsorb large amounts of water, natural gas, ions, or other substances. This property can enable shale to selectively and tenaciously hold or freely release fluids or ions.

Because shale has a tiny particle size, the interstitial spaces are very small. They are so small that oil, natural gas, and water have difficulty moving through the rock. Shale can, therefore, serve as a cap rock for oil and natural gas traps, and it is also an aquiclude that blocks or limits the flow of groundwater. Although the size of interstitial space in shale is very small, they can account for a significant volume of the rock. This volume allows the shale to hold substantial amounts of water, gas, or oil. However, shale cannot effectively transmit them because of the low permeability. The oil and gas industry overcame these limitations of shale by horizontal drilling and hydraulic fracturing techniques to create artificial porosity and permeability within the shale rock.

Measurement of Lithology and Mineralogy

In shale reservoirs, accurate petrophysical evaluation is difficult because of low porosity and permeability, variable and complex mineral composition, the presence of organic content, and acquisition problem of native state core samples. These conditions provide significant difficulty in the characterization of shale reservoir so that various methods were presented to evaluate lithology and mineralogy of shale reservoirs (Ahmed & Meehan, 2016; Ballard, 2007; Passey, Bohacs, Esch, Klimentidis, & Sinha, 2010; Quirein et al., 2010; Ramirez et al., 2011; Sondergeld, Newsham, Comisky, Rice, & Rai, 2010; Sondergeld & Rai, 1993). Determination methods for lithology and mineralogy in shale reservoirs include X-ray diffraction (XRD), Fourier transform infrared (FTIR) transmission spectroscopy, X-ray fluorescence (XRF), conventional log, elemental spectroscopy log, and cluster analysis.

Quantitative analysis of reservoir mineralogy is usually obtained from an evaluation of core samples, which are collected by side wall or conventional form or cuttings. Generally, XRD is commonly used to assess primary mineralogy. XRD, which is based on diffraction of X-rays by the crystal planes of minerals, measures the physical structure of a material (Ruessink & Harville, 1992). This method works sufficiently well but generally overpredicts quartz in clay-rich systems when clay separation is not performed (Sondergeld et al., 2010).

FTIR transmission spectroscopy, which relies on detection of molecular bond vibrations, measures the chemical composition of material (Ruessink & Harville, 1992). FTIR provides a fast and efficient way to eliminate this shortcoming without requiring clay fraction separation although organics must be removed before analysis. XRF is used by some laboratories to determine the mineralogy of clay-rich reservoirs (Ross & Bustin, 2006). XRF quantifies the elemental abundances that are then stoichiometrically apportioned to common minerals; excess carbon can be assigned to kerogen. XRF does not overpredict the amount of quartz (Sondergeld et al., 2010).

The conventional log also can determine lithology and mineral composition with supplementation of core analyses results such as XRD. A simplified measurement of lithology and mineralogy is performed by estimating the shale volume and cross-plotting multiple porosity responses in common with processes performed in conventional reservoirs. In unconventional shale reservoir, however, the amount of organic content should be determined before the process to extract the impact of the organic matters on the log responses (Ahmed & Meehan, 2016). Weight and volume percentages of kerogen can be acquired by conventional logs, such as density, GR, and Passey methods (Passey, Moretti, Kulla, Creaney, & Stroud, 1990), and core total organic carbon (TOC) measurements. Detail of kerogen measurements is presented in Organic Matters section. Conventional log measurements of GR, resistivity, density, sonic, neutron, and NMR can be input into the least squares error minimization solver. Results of core measurement can guide the selection of minerals to be input into the model.

The GR log is used to measure the formation of natural GR radioactivity. It measures the general GR emissions of all the radioactive elements such as potassium, uranium, and thorium together. Among the various sediments, shales present the strongest GR radiation so that the GR log is mainly used to quantify the shale volume. For most fine-grained rock evaluation, the GR is a critical well log to help differentiate shales from common reservoir lithologies, such as sandstone or carbonate. For shale gas plays, the source, seal, and reservoir are often entirely contained within the fine-grained rock lithofacies, and the GR curve may or may not be as useful as in conventional reservoirs. Generally, the potassium is included in the clays considerably. Although kaolinite and smectite show the very little amount of potassium, illite contains a large amount of potassium (Dresser Atlas, 1979). The clay mixtures with high kaolinite or high smectite content will have lower potassium radioactivity than clays made up principally of illite. Because most of clays are mixtures of several clay minerals, potassium radioactivity is observed in general shale reservoirs. The average content of potassium in shale reservoir is about 2%–3.5% (Rider, 2002). If shales are deposited under the marine condition, uranium forms unstable soluble salts. In this situation, uranium content has a positive relationship with organic matters so that the uranium content can be an indicator of organic substance (Fertl & Rieke, 1980). In lacustrine settings, there is generally a lack of uranium, and more often than not there is no relation between uranium and TOC (Bohacs, 1998; Bohacs & Miskell-Gerhardt, 1998). In these cases, the total GR curve remains an indicator of overall clay content in the rock (Bhuyan & Passey, 1994). In addition, it should be noted that the use of uranium is suitable for gas shale reservoirs that do not have uranium-enriched minerals such as apatite (Kochenov & Baturin, 2002).

The resistivity of a formation is directly related to electrically conductive components. The measurement of reservoir resistivity is of importance in logging because it is a method for identifying and quantifying hydrocarbons and water saturations. In conventional reservoirs, formation water is the primary conductor of electricity when the formation waters are brackish to saline, allowing for ionic conduction. Formation filled with saline water shows low resistivity. The larger the volume of saline water, the lower the resistivity of the rock filled with saline water. Inversely, hydrocarbons are nonconductive. When they are present in sufficient quantities, they displace the amount of water in a given formation. The rock sufficiently filled with hydrocarbons presents higher resistivity values than the rock filled with saline formation water (Archie, 1942). Many other factors affect the interpretation of resistivity, such as overburden pressure and pore pressure, temperature, and rock lithology and the percentage of conductive minerals.

Interestingly, some shale reservoirs show high conductivity, whereas others are not owing to the different depositional environment of the shale layers or different thermal history of the formation. Although it is usually expected that the resistivity of the formation will increase owing to the hydrocarbons and organic matters, it is correct only when the thermal maturity of the formation is high enough to generate hydrocarbons. Conversely, Anderson, Barber, Lüling, Sen, Taherian, Klein et al. (2008) showed that some shale layers could have high electric conductivity so that they show low resistivity. The cause of high conductivity is attributed to the presence of conductive minerals such

as pyrite or graphite. Pyrite is commonly present in organic-rich shale gas formations and may play a role in decreased resistivity response. In some shale gas reservoirs that are at very high maturities ($R_o \gg 3$), the formation resistivity can be lower than resistivity observed in the same formation at lower thermal maturities (R_o between 1 and 3). It was thought that the carbon in the organic contents recrystallizes to the mineral graphite (Passey et al., 2010). In extremely high-maturity organic-rich shale reservoirs ($R_o > 3$), the rock may be much more electrically conductive because of other mineral phases being present.

Cation-exchange capacity (CEC) of the clay minerals is another property that affects the resistivity of the shale layers. CEC value varies with the surface area of the clays. This means that the difference between the conductivity of clay species should be related to the surface area (Rider, 2002). Smectite has a larger specific surface area than the other clays and is, therefore, more conductive (Passey et al., 2010). The effect of CEC on shale conductivity depends on the salinity of the formation water. If the formation water salinity is greater than seawater salinity, the impact of excess conductivity due to clay minerals is small (Passey et al., 2010).

Neutron log measures the amount of hydrogen in a formation. Like other conventional well log interpretations, neutron log in gas shale layers is complex because most hydrogen in the organic matter, clay minerals, formation water, and hydrocarbons should be considered. Neutron log is affected not only by the hydrogen in the organic matter but also by the hydrogen in the hydroxyl (OH^-) in the clay minerals, as well as by the hydrogen in the formation of water and hydrocarbons (Passey et al., 2010). It is expected that neutron log response will be reduced in the gas shale layers because of the lower hydrogen of gas and organic matter compared with water. Owing to the increase in hydroxyl ions, this technique has limited application when the formation is not clay rich.

In shale reservoirs, elemental spectroscopy log has been used extensively to measure the lithology and mineral composition (Ahmed & Meehan, 2016). Elemental spectroscopy log tools record the energy spectra of the induced GRs to calculate the weight fractions of diverse elements in the formation. Neutrons released from a chemical source or pulsed neutron source emit induced GRs, and they interact with elements in the formation. Induced GRs are emitted from both the capture and inelastic energy spectra. Chemical source elemental spectroscopy log measures the formation elements in the capture energy spectrum, and pulsed neutron source elemental spectroscopy log measures the formation elements in both the capture and inelastic energy spectrum. The advantage of a pulsed neutron source elemental spectroscopy log is the ability to measure additional elements in the inelastic spectrum, mainly carbon. Even if some elements such as aluminum and magnesium are measurable in the capture energy spectrum, quantifying of these elements is problematic. The additional presence of these elements in the inelastic energy spectrum allows for a more accurate characterization of lithology and mineralogy (Pemper et al., 2006).

A spectral GR tool is also run in combination with the instrument to measure responses in the natural GR spectrum. These GRs are measured by a scintillation detector and processed to obtain elemental yields, and those are converted to elemental weight fractions (Pemper et al. 2006). A conversion for spectral deconvolution or inversion of the recorded GR spectra is applied to obtain relative elemental yields, which typically include aluminum, carbon, calcium, iron, gadolinium, hydrogen, potassium, magnesium, manganese, sodium, oxygen, sulfur, and silicon. Several service companies used individual spectroscopy instruments such as FLeX of Baker Hughes, and ECS and LithoScanner of Schlumberger to measure formation elements (Ahmed & Meehan, 2016).

Chemical source elemental spectroscopy log instruments measure the formation elements in the capture GR spectrum. The elemental weight fractions for silicon, calcium, magnesium, aluminum, iron, and sulfur may be input along with uranium, thorium, and potassium from spectral GR instruments and conventional log measurements of resistivity, density, neutron, and acoustic into the least squares error minimization solver. The selection of minerals to be input into the model is guided by core XRD analyses in the same shale play reservoir. Quirein et al. (2010) and Ramirez et al. (2011) presented the detailed workflow for establishing and predicting mineralogy, grain density, and porosity from chemical source elemental spectroscopy log. Core XRD bulk mineralogy data were used to determine the mineral model, and it was used as constraints in the log interpretation. Using chemical source logs, apparent volume, kerogen, and conventional logs such as porosity and resistivity, mineral bulk volumes were calculated simultaneously. The input for kerogen volume is obtained from basic regressions of core TOC weight fractions and conventional log measurements, such as density, uranium, GR, or the Passey method (Passey et al., 1990).

Pulse neutron source elemental spectroscopy logs have been used to measure the lithology and mineralogy of shale reservoirs. The capture and inelastic GR energy spectra from pulsed neutron source logging tools and natural GR from the conventional GR spectroscopy instrument are used to extract the chemistry of the subsurface formation being investigated (Jacobi et al., 2008; Pemper et al., 2006, 2009). The elemental concentrations measured in these methods include aluminum, calcium, iron, gadolinium, magnesium, sulfur, silicon, potassium, thorium, uranium, titanium, and carbon. The strength of pulse neutron source elemental spectroscopy log is a measurement of carbon. Inelastic spectrum generated by pulsed neutron source enables to measure carbon. Neutrons emitted from the instrument interact with formation elements so that capture and inelastic energy spectrums are emitted. The GRs from each spectrum are separated based on the characteristic GR emission of the individual component, and then algorithms are used to convert these yields into concentrations. Pemper et al. (2006) provided diagnostic methods. It begins with a broad evaluation of the general lithology, followed by a more detailed assessment of the specific lithology, ultimately leading to a determination of the mineralogic content of the formation. After this is determined, the amount of carbon can be identified as organic carbon. Wang and Carr (2013) constructed a three-dimensional (3-D) lithofacies model for the Marcellus Shale with the workflow shown in Fig. 2.1. They integrated core data, pulsed neutron spectroscopy logs, conventional logs, and seismic data as well as regional geologic knowledge.

Lithology and Mineralogy in Several Shale Plays

Although completion treatments in Barnett Shale with horizontal drilling and hydraulic fracturing have been successful in North America, the same methods did not succeed in all other shale formations. As demonstrated by several previous studies, all shale plays are not clones of the Barnett Shale, and they show extremely various lithology and mineralogy even within the single shale play (He et al., 2016; Rickman, Mullen, Petre, Grieser, & Kundert, 2008; Rutter, Mecklenburgh, & Taylor, 2017; Sone & Zoback, 2013). Mineralogic characteristics affect significantly on the petrophysical and geomechanical properties, which considerably affect productivity in shale reservoirs. The dominant mineral components in shales are quartz, feldspar, pyrite, various clay minerals such as montmorillonite, illite, smectite, and kaolinite, and carbonate minerals in varying proportions. Fig. 2.2 presents the ternary diagram with dominant mineral contents in several shale

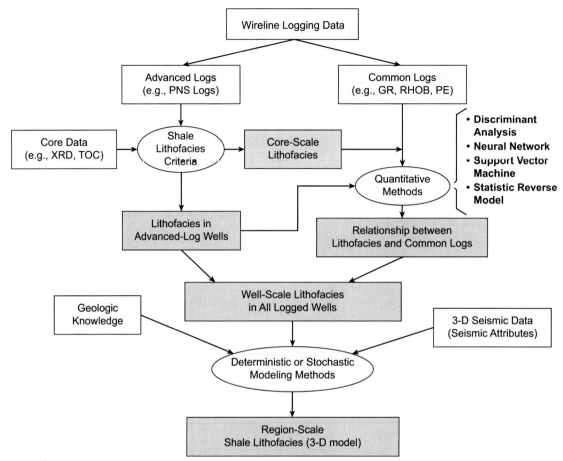

FIG. 2.1 The workflow to construct a three-dimensional (3-D) lithofacies model for the Marcellus Shale of the Appalachian Basin. (Credit: Wang, G., & Carr, T.R. (2013). Organic-rich marcellus shale lithofacies modeling and distribution pattern analysis in the Appalachian basin organic-rich shale lithofacies modeling, appalachian basin. *AAPG Bulletin, 97*(12), 2173–2205. https://doi.org/10.1306/05141312135.)

reservoirs (Morley et al. 2018). As shown in Fig. 2.2, various shale plays present extensive types of mineral compositions. In the following section, lithology and mineralogy of several shale plays are briefly introduced.

The Barnett Shale is present across the Fort Worth Basin and adjoining Bend Arch in north-central Texas, extending over a total area of 28,000 square miles (NETL, 2011). The shale play covers approximately approaching 7000 square miles, which is a fourth of the entire geographic area of the Barnett and roughly 4000 square miles have been currently activated (Ahmed & Meehan, 2016). The depth is 6500 to 8500 ft with the gross thickness increasing from 100 to 600 ft through west to east. NETL (2011) reported detailed geology of Barnett Shale including lithology and mineralogy. Depending on several studies for the formation lithology, Barnett Shale consists of black siliceous shale, limestone, and minor dolomite (Loucks & Ruppel, 2007; Montgomery, Jarvie, Bowker, & Pollastro, 2005; NETL, 2011; Papazis, 2005). Bowker (2003) reported that the mineral composition with 45% of quartz, 27% of illite with minor smectite, 8% of calcite and dolomite, 7% of feldspars, 5% of pyrite, 3% of siderite, and the slightest amount of native copper and phosphate material. Givens and Zhao (2004) presented a similar mineral composition except for considerably higher calcite and dolomite, which is 15%–19%. Jarvie, Hill, Ruble, and Pollastro (2007) measured mineral contents in a Barnett well with 40%–60% of quartz, 40%–60% of clay minerals, and a variable

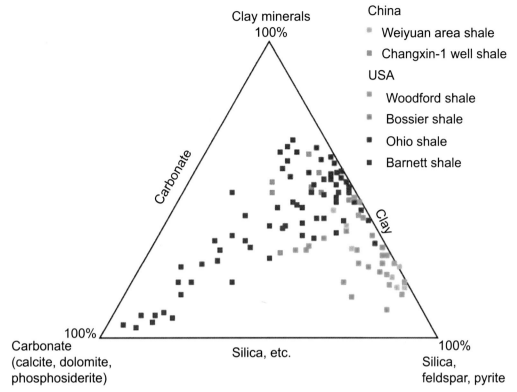

FIG. 2.2 Ternary diagram of several shale gas reservoirs. (Credit: Morley, C. K., von Hagke, C., Hansberry, R., Collins, A., Kanitpanyacharoen, W., & King, R. (2018). Review of major shale-dominated detachment and thrust characteristics in the diagenetic zone: Part II, rock mechanics and microscopic scale. *Earth-science Reviews*, *176*, 19–50. https://doi.org/10.1016/j.earscirev.2017.09.015.)

calcite content. Loucks and Ruppel (2007) reported mineralogy for siliceous-mudstone lithofacies with 41% of quartz, 29% of clays, 9% of pyrite, 8% of feldspar, 6% of calcite, 4% of dolomite, and 3% of phosphate. Based on these various studies of mineralogy, the mean mineral composition of Barnett Shale is 35%–50% of quartz, 10%–50% of clay minerals primarily with illite, 0%–30% of calcite, dolomite, and siderite, 7% of feldspars, 5% of pyrite, and trace of phosphate and gypsum (NETL, 2011). Jarvie et al. (2007) presented that the relative brittleness of shale can be assessed by the ratio of quartz to the sum of quartz, calcite, and clay minerals. In other words, the formation of Barnett Shale, which is dominant with siliceous shale, is brittle, and has a suitable condition for performing hydraulic fracturing technique.

The Middle Devonian Marcellus Shale is located through the central Appalachian Basin extending for roughly 600 miles. Marcellus Shale is presented in Ontario, New York, Pennsylvania, Ohio, West Virginia,

Maryland, and Virginia with a total area of almost 75,000 square miles (NETL, 2011). Various reports were also presented for lithology and mineral composition for Marcellus Shale (Harper, 1999; Larese & Heald, 1976; Milici & Swezey, 2006; Nuhfer, Vinopal, & Klanderman, 1979; Potter, Maynard, & Pryor, 1980; Roen, 1993; Wrightstone, 2009). Geologically, Marcellus Shale is similar to Barnett Shale. Lithology and mineral composition are also analogous. Typical mineral composition of Marcellus Shale is 10%–60% of quartz, 10%–35% of clay minerals primarily with illite, 3%–50% of calcite, dolomite, and siderite, 0%–4% of feldspars, 5%–13% of pyrite, 5%–30% of mica, and trace of phosphate and gypsum (NETL, 2011). Both Marcellus and Barnett plays show significant success because of their siliceous shale, which behaves in a brittle fashion so that it is prone to hydraulic fracturing.

As shown in Fig. 2.3, there are various current-producing and prospective shale plays in the United States (EIA, 2016). Including Barnett and Marcellus,

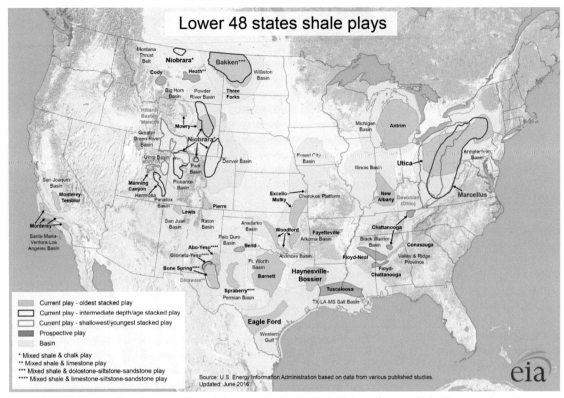

FIG. 2.3 Current and prospective shale plays in the United States. (Source: U.S. Energy Information Administration (Jun 2016).)

Fayetteville, Woodford, Haynesville, and Eagle Ford plays are known as the Big Six Shale in the United States (Ahmed & Meehan, 2016). The Fayetteville is geologically equivalent to the Barnett shale characteristics. Lithology of the Fayetteville is termed as siliceous shale with 20%–60% of silica, with very little carbonate and clay. Woodford Shale presents a dark gray to black shale, which has a high TOC roughly up to 9.8%. Woodford Shale is composed of 50%–65% of silica, 5%–10% of carbonate, and 30%–35% clay. Lithology of the Haynesville shale, generally termed siliceous marl, consists of 25%–45% of silica, 15%–40% of carbonate, and 30%–45% of clay, which makes it a more ductile formation and difficult to hydraulic fracturing. Eagle Ford Shale is composed of organic-rich calcareous mudrock with mineral composition ranging from 10% to 25% of silica, 60%–80% carbonate, and 10%–20% clay.

ORGANIC MATTERS

The conventional petroleum system is composed of source rock, reservoir rock, trap, and seal. In a typical source rock, some of the hydrocarbons are expelled and migrate into reservoir rock to become conventional reservoirs. In unconventional shale gas and oil reservoirs, a substantial volume of the generated hydrocarbon cannot be expelled from the source rock, and the source rock itself becomes the reservoir. The most important characteristic of a source rock is that it contains a high amount of organic matters. Organic matters, which has the capacity to produce and store hydrocarbons, play an essential role in the unconventional petroleum system of shale. TOC consists of three components such as the gas or oil that is already present in the rock, the kerogen that represents the available carbon that could be generated, and residual carbon that has no potential to create hydrocarbons (Jarvie, 1991).

Hydrocarbon is formed within the source rock from kerogen and bitumen when they are subjected to increasing temperature and pressure. Kerogen is a solid mixture of organic chemical compounds that are insoluble in usual organic solvents because of the vast molecular weight of its component, and bitumen is a solid or nearly solid inflammable organic matter generated from kerogens such as asphalt and mineral wax.

The quantity of organic carbon is measured as the TOC content of rocks. Expressed generally by the weight percent of organic carbon, TOC is the concentration of organic matters in the rocks. The volume percent of TOC is approximately twice larger than the weight percent. As a conventional effective source rock, a value of about 0.5% TOC by weight percent is considered a minimum or threshold. For shale reservoirs, values of about 2% are considered as a minimum and may exceed 10%−12%. According to Dayal and Mani (2017), hydrocarbon generation potentials are evaluated as the following: very poor: <0.5%; poor: 0.5%−1%; fair: 1%−2%; good: 2%−4%; very good: 4%−12%; and excellent: >12%. For commercial shale gas production, several studies suggested a target of TOC at least 2%−3% as a lower limit in the United States (Lu et al., 2012). Typical TOC values for the Barnett and Marcellus shales are 2%−6% and 2%−10%, respectively. Depending on the examination of shale gas formations worldwide by Passey et al. (2010), the total porosity and gas content is directly associated with the TOC content of the rock (Fig. 2.4). Fig. 2.4 presents that high local TOC is a critical factor to evaluate the potential of shale gas reservoirs because it relates to both porosity and gas saturation.

As mentioned earlier, organic-rich shales show considerable compositional variability (Passey et al., 2010; Plint, Macquaker, & Varban, 2012; Quirein et al., 2010; Ramirez et al., 2011; Sondergeld et al., 2010; Trabucho-Alexandre, Hay, & De Boer, 2012). It is because of the diverse and dynamic nature of the depositional processes and environments of shale formations. The composition, structures, and organic contents of shale reservoirs rely on the physical, chemical, and biological depositional processes and their depositional environments. The depositional processes affect grain assemblages, mineralogy, and types of deposited kerogen. Unconventional shales can be generated in various environments from the bottom of lakes to the plains of the deep ocean (Schieber, 2011; Stow, Huc, & Bertrand, 2001; Trabucho-Alexandre et al., 2012).

Type of Kerogen

Type of kerogen is generally classified into four primary categories (Tissot & Welte, 1984) or seven specific types currently used in the hydrocarbon industry (Ahmed & Meehan, 2016). Kerogen types of I, II, III, and IV are presented by Tissot and Welte (1984), and then kerogen types of IS, IIS, IIIC, and IIIV have been additionally considered in these days. These types of kerogens are based on the kerogen character, elemental contents, and depositional environments. Original hydrogen and oxygen contents and organic-type origin can be distinguished by kerogen types. Investigation of these kerogen types is critical to understand the processes of storage, retention, and release of hydrocarbons.

Type I kerogen is mainly generated from the lacustrine depositional environment. It has the original content of highest hydrogen and lowest oxygen from a microorganism source, such as algae, plankton, and other organic matters that have been reworked by bacteria. Type II kerogen is normally formed in anoxic marine or the transitional marine depositional environment and is characterized as being hydrogen rich and oxygen deficient. Type II kerogen is derived from mixtures of algal, plankton, and other organic matters from bacteria along with structured kerogen plant materials. Type IS and IIS kerogens are generated in depositional environments with increased sulfur compounds in the kerogen. The generation of oil in Type IS and IIS kerogens starts much earlier because of kinetic reactions involving sulfur-containing compounds. Source rocks including Types I, IS, II, and IIS kerogens are prone to generate liquid hydrocarbons mainly.

Types IIIC, IIIV and Type IV kerogens are defined as having debris materials from structured woody plant origin that were originally deposited within terrestrial or transitional marine environments, such as swamps, delta complexes, or shallow lagoons. These kerogen types exhibit much lower original hydrogen content and higher oxygen content. Type IIIC kerogens are most often assigned to a transitional marine depositional system where hydrogen-rich algal microorganisms have been preserved. Type IIIV is defined as dominantly structured plant vitrain macerals that are most often gas prone. Type IV kerogen originates chiefly from residual organic matters so that it is incapable of generating hydrocarbons. These kerogens are either kerogens that are fully transformed by weathering, combustion, and biologic oxidation or charcoal materials that are assigned as burnt plant materials. This type of kerogen has almost no potential for hydrocarbons (Hunt, 1996; Tissot & Welte, 1984).

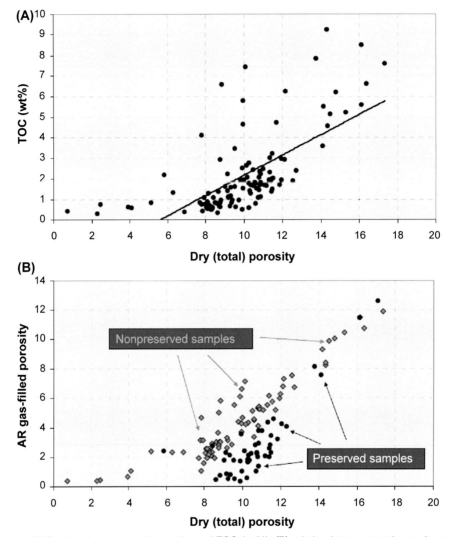

FIG. 2.4 **(A)** Relation between total porosity and TOC (wt%); **(B)** relation between total porosity and as-received gas-filled porosity for preserved and nonpreserved samples. (Credit: Passey, Q.R., Bohacs, K., Esch, W.L., Klimentidis, R., & Sinha, S. (2010). From oil-prone source rock to gas-producing shale reservoir — geologic and petrophysical characterization of unconventional shale gas reservoirs. *Paper presented at the international oil and gas conference and exhibition in China, Beijing, China.* https://doi.org/10.2118/131350-MS.)

Kerogen types can be illustrated within the Van Krevelen diagram (Fig. 2.5). This plot includes the defined relationship of rock pyrolysis originated hydrogen index (HI) versus oxygen index (OI) to original TOC (TOC_o). This plot is also useful in defining the transformation of specific kerogen types with greater maturity. This helps explain how present-day TOC (TOC_{pd}) is different from the original TOC. Four primary kerogen types can be quickly assessed in this plot of HI versus

OI. It is easy to use HI and OI values due to Rock-Eval pyrolysis analyses, which are less expensive and quicker to obtain than the atomic H/C versus O/C analysis (Tene, Bosma, Al Kobaisi, & Hajibeygi, 2017; Peters & Cassa, 1994).

Thermal Maturity

With the identification of TOC content and type of kerogen, thermal maturity is another essential character

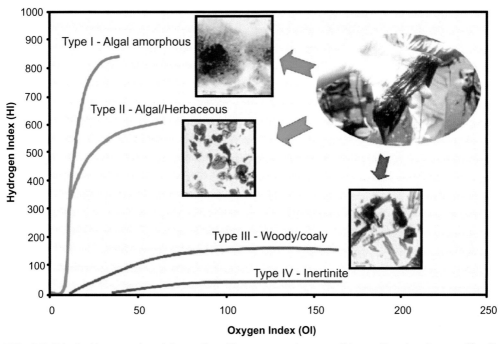

FIG. 2.5 Principal types and evolution paths of kerogen are shown on this van Krevelen diagram. (Credit: Passey, Q.R., Bohacs, K., Esch, W.L., Klimentidis, R., & Sinha, S. (2010). From oil-prone source rock to gas-producing shale reservoir — geologic and petrophysical characterization of unconventional shale gas reservoirs. *Paper presented at the international oil and gas conference and exhibition in China, Beijing, China.* https://doi.org/10.2118/131350-MS.)

to understand the overall petroleum generation potential. Thermal maturity is an indication of the maximum temperature exposure of sediment and extent of temperature-time-driven reactions (Ahmed & Meehan, 2016; Bust, Majid, Oletu, & Worthington, 2011; Dayal & Mani, 2017; Rezaee, 2015). In other words, thermal maturity is the effectiveness of the organic sources. It is used to describe the thermal degradation and conversion of sedimentary organic contents to oil and gas as being in the petroleum window. In cases of reservoirs that are Type I and II kerogen dominant, kerogens transform into liquid hydrocarbons within a phase that is termed catagenesis. Through attaining the required activation energy levels, these kerogen types develop gases and an intermediate liquid hydrocarbon phase that is termed bitumen. Bitumen then transforms into liquid petroleum hydrocarbons. With higher levels of time, pressure, and temperature, the metagenesis phase involves cracking these generated liquid hydrocarbons and gases into thermally stable hydrocarbon gases of propane, ethane, and methane. With even higher levels of time, temperature, and pressure, the secondary

cracking phase will also transform these thermally stable gas types into methane gas. In Type III and IV kerogen-dominant reservoirs, the sequence of primary hydrocarbon generation is simple that kerogens convert to a minor amount of hydrocarbon fluids and then finally to pure methane gas.

Thermal maturity generally can be quantified by vitrinite reflectance ($\%R_o$), which is an indication of the maximum paleo-temperature exposure. Vitrinite is an organoclast, fossilized wood particle altered from woody plant tissues (Rezaee, 2015). When such particles are identified in kerogen, a light beam is shined onto the polished surface, and the amount of light reflected recorded by a photomultiplier. Thermally immature source rocks, which present lower than $0.6\%R_o$ of vitrinite reflectance, do not have a pronounced effect of temperature (Mani, Patil, & Dayal, 2015). Thermally mature organic matters, which present $0.6\%-1.35\%R_o$ of vitrinite reflectance, generates oil. The postmature organic matters, which show higher than $1.5\%R_o$ of vitrinite reflectance, is in wet and dry gas zones (Mani et al., 2015; Tissot & Welte, 1984). In addition, Rock-

Eval pyrolysis temperature (T_{max}) is also popularly used to evaluate the thermal maturity of kerogen. T_{max} value is the temperature at which the maximum amount of hydrocarbons is generated from the decomposition and conversion of kerogen. The T_{max} values under 435°C specify an immature kerogen for generation of hydrocarbons, the temperature range between 435 and 465°C suggests a mature stage, and those more than 465°C show a postmature stage, suitable for generation of gas (Hunt, 1996; Mani et al., 2015; Tissot & Welte, 1984).

PORE GEOMETRY

Previous studies presented that various depositional processes and sources of sediment supply affect grain accumulations, lithology, mineral composition, and deposited kerogen types (Ko, Loucks, Ruppel, Zhang, & Peng, 2017; Macquaker, Keller, & Davies, 2010; Myrow & Southard, 1996; Schieber, Southard, & Kevin, 2007; Young & Southard, 1978). Furthermore, in conjunction with maturity and diagenesis, they contribute to the complexity and heterogeneity of shale rock pore types and morphology, which directly affect hydrocarbon productivity. In unconventional shale reservoirs, intricate pore network system enormously contributes to complex flow mechanisms. Pore networks of shale plays are comprised of organic matters and inorganic materials as well as natural fractures. Generally, pores observed in shale reservoirs show nanometer to micrometer size range. The complexity of pore geometry and microscale heterogeneity can considerably impact matrix permeability and flow behaviors in shale reservoirs (Ko et al., 2017). Understanding complex pore geometry is fundamental for defining storage capacity and fluid flow models of shale reservoirs.

Classification of Pore Types

Pores in shale reservoirs are classified as inorganic mineral pores and organic matter pores (Ko, Zhang, Loucks, Ruppel, & Shao, 2015, Ko, Zhang, Loucks, Ruppel, & Shao, 2017; Loucks, Reed, Ruppel, & Hammes, 2012; Pommer & Milliken, 2015). The inorganic mineral pores can be subdivided into primary interparticle or intraparticle mineral pores originally saturated with formation water and modified mineral pores containing migrated petroleum such as bitumen and residual oil. The organic matter pores comprise three categories based on their morphology and interpreted origins. The organic matter pores can be divided into primary organic matter pores and secondary organic matter pores. Secondary organic matter pores also can be subdivided into organic matter bubble and spongy pores. Primary organic matter pores are related to the original structure of kerogen, and organic matter bubble pores and spongy pores are the results of the maturation of kerogen (Ko et al., 2015). Generally, the size of organic matter bubble pores is larger, and the amount of them is lower than organic matter spongy pores in shale reservoirs (Ko et al., 2017).

Primary mineral pores include interparticle pores and intraparticle pores. The interparticle pores are defined as pores which occur between grains and crystals. The interparticle pores are commonly well connected and form an effective pore network that can contribute to permeability. The intraparticle pores are generated within grain and crystal boundaries. As a result of clay mineral deformation, some of the pores within clay mineral platelets can be generated. Clay minerals are soft and ductile so that they are easily bent around rigid grains during compaction. Their origin commonly controls the shape and size of the intraparticle pores. The intraparticle pores are relatively isolated and normally cannot contribute to permeability. Fig. 2.6A and B illustrate intraparticle pores between cemented coccolith hash and elongated intraparticle pores within clay platelet, respectively (Ko et al., 2017).

To avoid misinterpretation and misidentification as organic matter pores, Ko et al. (2017) separated modified primary mineral pores from interparticle and intraparticle pores. Modified primary mineral pores are in contact with organic matters so that it is crucial to recognize modified primary mineral pores correctly when quantifying pore types. Two processes can generate modified primary mineral pores. First, oil and gas leave the original mineral pore, and then residual oil is formed when heavy hydrocarbon component is adsorbed on the mineral surface. The shape and size of modified primary mineral pores depend on the pore formed by the original mineral pores or the surrounding grains. Second, modified primary mineral pores can also be generated by migration of petroleum and infill of surrounding mineral pores, which include early gas or connate water on the edge of minerals. In this case, the pores are in contact with both minerals and organic matters. The size of modified primary mineral pores ranges generally from a few micrometers to tens of micrometers in diameter, but some of the pores show nanometer scale. Although most of the pores have irregular shapes, some pores are rounded. Fig. 2.6C shows modified primary mineral pores in Eagle Ford Shale (Ko et al., 2017).

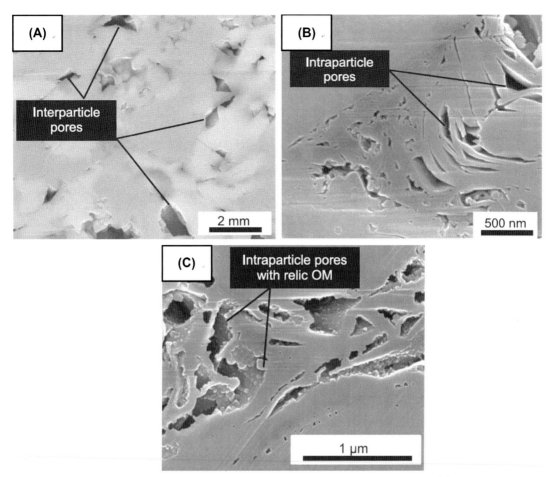

FIG. 2.6 Scanning electron microscope photomicrographs of three different pore types: **(A)** interparticle, **(B)** intraparticle, and **(C)** modified primary mineral pores in Eagle Ford samples. (Credit: Ko, L.T., Loucks, R.G., Ruppel, S.C., Zhang, T., & Peng, S. (2017). Origin and characterization of Eagle Ford pore networks in the south Texas Upper Cretaceous shelf. *AAPG Bulletin, 101*(3), 387−418. https://doi.org/10.1306/08051616035.)

Primary organic matter pores include original kerogens such as cells in wood and leaf fragments or spores. These pores are normally particulate and noncompactible owing to rigid cell structures. Primary organic pores, which are originally inherited from kerogen, are not related to thermal maturation. Secondary organic matter bubble and spongy pores are related to thermal maturation. Organic matter bubble pores are generated from bitumen cracking to hydrocarbon liquids (Loucks & Reed, 2014). Bubble pores are commonly rounded, and their size ranges from hundreds of nanometers to a few micrometers. Organic matter spongy pores are generated from high maturation stage in which gas is created (Loucks et al., 2012). Therefore, spongy pores are generally found in migrated solid bitumen,

pyrobitumen, or char (Bernard, Wirth, Schreiber, Schulz, & Horsfield, 2012a; Bernard, Wirth, Schreiber, Schulz, & Horsfield, 2012b; Loucks & Reed, 2014). Organic matter spongy pores have various pore shapes ranging commonly from 2.5 to 200 nm in size. Generally, organic matter spongy pores are smaller and much more abundant than organic matter bubble pores in shale reservoirs (Ko et al., 2017). In addition, there is a special case of organic matter pores depending on the pore location rather than origin. Organic matter pores in kerogen and solid-bitumen complex are special but common in shale reservoirs. Organic matter spongy can be generated in a complex of kerogen and bitumen, a solid bitumen or pyrobitumen, which is trapped within kerogen or adsorbed on the kerogen surfaces.

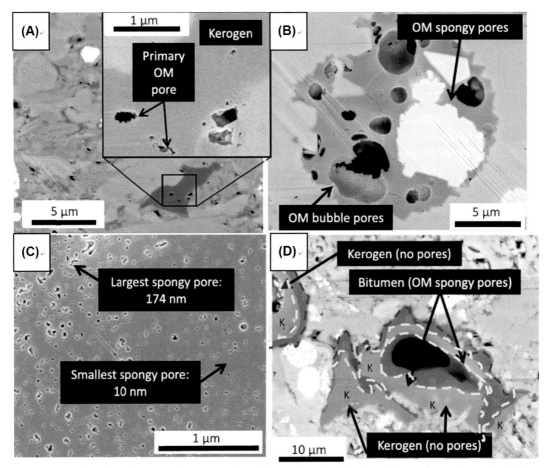

FIG. 2.7 Scanning electron microscope photomicrographs showing three types of organic matter: **(A)** primary organic matter pores, **(B)** organic matter bubble pores, **(C)** organic matter spongy pores, and **(D)** organic matter pores in kerogen and solid-bitumen complex. B = bitumen (*dark gray*), K = kerogen (*yellow*)). (Credit: Ko, L.T., Loucks, R.G, Ruppel, S.C, Zhang, T., & Peng, S. (2017). Origin and characterization of Eagle Ford pore networks in the south Texas upper cretaceous shelf. *AAPG Bulletin, 101*(3), 387–418. https://doi. org/10.1306/08051616035.)

Fig. 2.7 presents the primary organic matter pores in compacted kerogen, the organic matter bubble pores in bitumen within the globigerinid chamber, the organic matter spongy pores in the bitumen inside a foraminifera chamber, and the organic matter pores in a complex of kerogen and bitumen complex.

Pore Network Evolution

Pores within organic-rich shale reservoirs have become a subject of increased attention so that the nature of pore and pore network evolution in shale reservoir are documented by several studies (Ambrose, Hartman, Diaz Campos, Akkutlu, & Sondergeld, 2010; Bernard et al., 2012a; Curtis, Ambrose, & Sondergeld, 2010; Fishman et al., 2012; Houben, Desbois, & Urai, 2014; Kuila & Prasad, 2011, Kuila & Prasad, 2013; Loucks et al., 2012, Loucks et al., 2009; Pommer & Milliken, 2015). These researches presented heterogeneity in the character of pore network in shale reservoirs and model of pore network evolution which is controlled by both the composition of grain accumulations and burial conditions.

In early, uncompacted sediment statement, fine-grained sediments show large pore volumes up to 80% (Velde, 1996). In this stage, primary mineral inter-particle and intraparticle pores can be generated, and they are prone to compaction with some early cementation (Desbois, Urai, & Kukla, 2009; Loucks et al., 2012; Milliken, Ko, Pommer, & Marsaglia, 2014; Milliken and Reed, 2010). During early compaction, much of the primary pore space is destroyed, especially within abundant ductile components such as marine kerogen (Loucks et al., 2012; Milliken et al., 2014; Mondol et al., 2007; Pommer, 2014; Velde, 1996). In this early burial stage, large and rigid grains protect large pores from compaction so that they preserve primary pore space (Desbois et al., 2009; Milliken et al., 2014; Milliken and Reed, 2010; Pommer, 2014). For example, Eagle Ford Shale presents a positive correlation between calcite abundance and pore volume (Pommer & Milliken, 2015). In this case, rigid skeletal calcite grains resist to compaction so that they shelter porosity between grains. On the contrary, Eagle Ford Shale shows negative correlation between organic matter abundance and pore volume, and it suggests that detrital organic matters behave in a ductile manner. Organic matters are particularly sensitive to compaction during early diagenesis. In both early sediment and burial stage, organic matter pores exist only in a trace amount in kerogen (Fishman et al., 2012; Milliken et al., 2014). Minerals precipitated from solution in diagenesis processes may occlude pore spaces. Calcite, quartz, pyrite, kaolinite, and apatite are most commonly generated as pore-filling cement as well as replacements of detrital grains.

Fig. 2.8 provides a common diagenetic pathway of Eagle Ford sediments (Pommer & Milliken, 2015). In the low-maturity situation, mineral-associated pores are dominated with a trace of primary organic matter pores in pore networks. However, because of compaction and infill of pore space by organic matters such as bitumen, inorganic mineral pores, primary organic matter pores are absent in high-maturity condition. In high-maturity formations that represent the gas and condensate window, porosity is well developed within organic matters due to later burial diagenetic processes (Fishman, Guthrie, & Honarpour, 2013; Pommer & Milliken, 2015; Reed & Ruppel, 2012). Substantial volumes of secondary organic matter pores are formed as volatiles are generated and expelled. Thermal maturity has the greatest impact on the development of organic matter pores, and thus, on the overall pore networks of shale reservoirs (Fishman et al., 2013).

Pore Size Classification

In unconventional shale reservoirs, nanometer to micrometer size of pores form flow networks in the matrix, and they significantly contribute to oil and gas flows. Generally, pore size in shale reservoir is less than 1 µm. Understanding of detailed pore size distribution can help accurate characterization of reservoirs. International Union of Pure and Applied Chemistry (IUPAC) presented pore size terminology and classification standard (Rouquerol et al., 1994). Rouquerol et al. (1994) defined three distinct pore size distributions: micropores, which have width less than 2 nm; mesopores, which have width between 2 and 50 nm; and macropores, which have width greater than 50 nm as shown in Fig. 2.9. However, with this classification, nearly all mudrock pores would be grouped with larger pores in carbonates and sandstones as macropores. Although the pore size classification of Rouquerol et al. (1994) is appropriate for chemical products, it would be an insufficient standard for reservoir systems, especially in shale reservoirs. Modifying the pore size classification of Choquette and Pray (1970), Loucks et al. (2012) provided a useful pore classification for the reservoir. Loucks et al. (2012) presented five distinct pore size distributions such as picopores, which have width less than 1 nm; nanopores, which have width between 1 nm and 1 µm; micropores, which have width between 1 and 62.5 µm; mesopores, which have width between 62.5 µm and 4 mm; and macropores, which have width larger than 4 mm (Fig. 2.9).

In nanoscale pores in an unconventional shale reservoir, thermodynamic phase behavior of oil and gas are slightly different from the conventional pore system. According to several studies, spatial confinement in nanopores changes the phase behavior and fluid properties due to pore-proximity effects (Alharthy, Weldu Teklu, Nguyen, Kazemi, & Graves, 2016; Barsotti, Tan, Saraji, Piri, & Chen, 2016; Chen, Mehmani, Li, Georgi, & Jin, 2013). Because of nanoscale pore size, van der Waals intermolecular forces, which are ignored in the conventional system, should be considered. To consider this effect in confined pore system, Alharthy et al. (2016) suggested another pore size classification. They presented three pore size distributions: confined pores, which have width between 2 and 3 nm; midconfined pores, which have width between 3 and 25 nm; and unconfined pores, which have width larger than 1000 nm. To consider accurate fluid behavior in nanopore system, confinement pore system should be clearly understood. Details of phase behavior mechanisms in nanopores are presented in Chapter 3 of this book.

FRACTURE SYSTEM

Fracture is a discontinuity or parting caused by a brittle failure in a reservoir (Narr, Schechter, & Thompson,

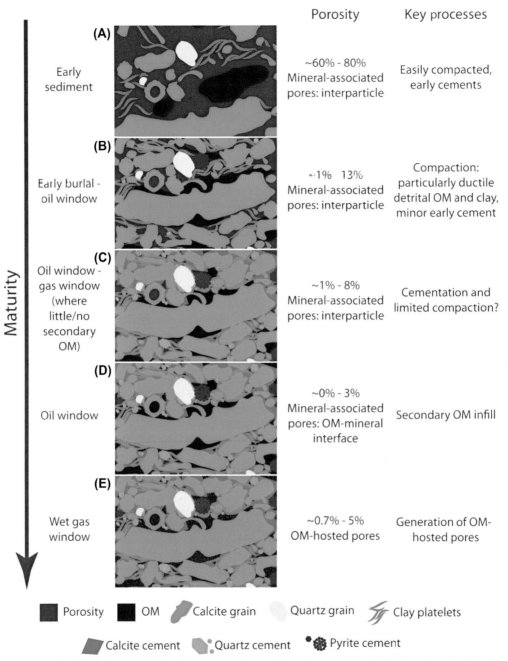

FIG. 2.8 Simplified diagram displaying common diagenetic pathways of Eagle Ford sediments. (Credit: Pommer, M., & Milliken, K. (2015). Pore types and pore-size distributions across thermal maturity, Eagle Ford formation, southern Texas Pores across thermal maturity, Eagle Ford. *AAPG Bulletin, 99*(9), 1713—1744. https://doi.org/10.1306/03051514151.)

2006). In the Earth's crust, there are extremely numerous natural fracture systems that are caused by tectonic forces in formations. Generally, there are three types of principal natural fractures: join, fault, and contractional fracture. These types of natural fractures show separate features such as origins, occurrences,

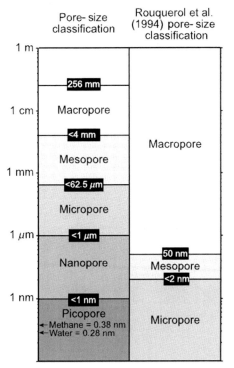

FIG. 2.9 Pore size classifications for mudrock pores, which are defined by Loucks et al. (2012) and Rouqueraol et al. (1994) of IUPAC, respectively. (Credit: Loucks, R. G., Reed, R., Ruppel, S.C., and Hammes, U. (2012). Spectrum of pore types and networks in mudrocks and a descriptive classification for matrix-related mudrock pores. *AAPG Bulletin, 96*(6):1071−1098.)

the direction of the Earth's stress field. In most petroleum reservoirs, the stress field varies in direction and magnitude with location. Therefore, understanding complex natural fracture systems is significantly crucial for effective hydrocarbon production.

In most productive unconventional shale reservoirs, natural fractures are present. Natural fractures have been considered as a critical factor for high productivity in shale reservoirs with low porosity and low permeability. Natural fractures can be classified into three types based on seismic responses: open, partially open, and mineralized fractures, which indicate conductive, mixed, and resistive fractures, respectively, in a natural state (Ahmed & Meehan, 2016). In production from shale reservoir, even partially open and mineralized fractures can positively affect hydrocarbon productivity by reactivation of hydraulic fracture stimulation. Natural fractures combined with the hydraulic fractures can generate a complex fracture network (Gale, Laubach, OlsonEichhubl, & Fall, 2014). Therefore, understanding the natural fracture system is critical for successful completion in shale reservoirs. In addition, during reservoir transition to an enhancement process such as CO_2 injection, natural fracture system can become paramount. In the enhancement process, natural fractures, which are ineffective under primary production process, can be important because of the strong reactivation effect. As shown in Section 5.1, natural fracture systems can be a major factor for CO_2 injection because the connectivity of horizontal wells depends on natural fractures. In shale reservoirs, natural fractures improve the movement of fluids, increase the conductivity, and affect fluid recovery efficiency.

In the petroleum industry, several methods have been used to simulate the natural fracture system: single effective medium, dual porosity and dual permeability, and discrete fracture models. The single effective medium model uses the properties whose matrix and fractures are combined to generate a single effective reservoir continuum (Narr et al., 2006). In this model, the fluid interaction between matrix and fracture is considered by pseudorelative permeability functions. Although single effective medium model is simple, it cannot simulate complex matrix-fracture networks. It is appropriate only where reservoir performance is dominated mainly by fluid behavior rather than matrix-fracture interactions and where fluid flow in the reservoir is single phase. Being initially presented by Warren and Root (1963), dual porosity model has been commonly used for the natural fracture system in the oil industry. In the dual porosity model, fractures are considered mainly as a flow path of hydrocarbon,

and characteristics so that the impacts of them on reservoir fluid flow are also different. The most significant discrimination between joint and fault is the existence of shear displacement. Joint, or extension fracture, is formed only with tension. Fracture walls are pulled perpendicularly away from each other, and there is no shear displacement during this process. On the contrary, the fault is generated from shear movements parallel to the fracture plane and at angles that range from parallel to perpendicular to the fracture propagation orientation (Rezaee, 2015). Fig. 2.10 shows general fracture style including joint, normal fault, reverse or thrust fault, and strike-slip fault (Narr et al., 2006). Contractional fractures can be generated by contractional displacements that are caused by volume loss across a plane. The volume loss can result from crushing (deformation bands), grain rearrangement (compaction bands), or by chemical dissolution (stylolites). The orientation of natural fractures is dominated by

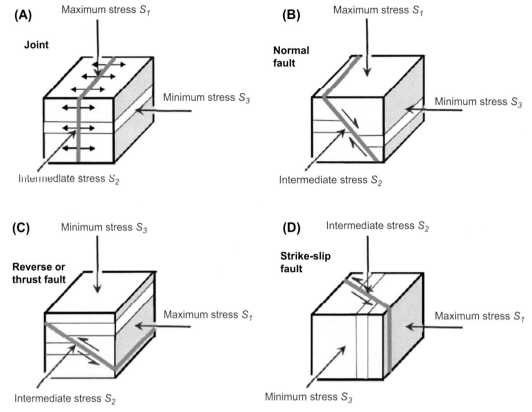

FIG. 2.10 General fracture styles with their displacements and orientations relative to principal stress orientations. (Credit: Narr, W., Schechter, D.S., & Thompson, L.B. (2006). *Naturally fractured reservoir characterization*: Society of Petroleum Engineers.)

whereas the matrix is regarded as hydrocarbon storage room. Matrix and fractures are separated into discrete continua which have respective hydraulic properties. Although a dual porosity model requires more detailed data and computational cost than the single effective medium model, it is not immense problems these days. The more critical problem of dual porosity model is the enormously various configurations of fracture systems. Shale reservoirs present distinct and unique fracture characteristics. Dual porosity model, which excessively simplifies complex fracture system, can cause a misunderstanding of overall reservoir process. To consider specific characteristics of the fracture system, discrete fracture model can be applied. Discrete fracture model considers the geometry of a fully defined individual fracture and then simulates fluid flows in a generated complex fracture system. Discrete fracture model presents the most geologically realistic approach

to model the flows of fluid in a fracture system. However, discrete fracture model requires significantly longer computational time and more detailed data than dual porosity model so that it cannot be applied for full-field unconventional shale reservoir simulation yet.

Dual Porosity Model

In the petroleum industry, naturally fractured reservoirs are generally characterized by dual porosity system. Barenblatt, Zheltov, and Kochina (1960) first introduced the concept of a dual porosity model, which presents two distinctive porous regions with different properties. The first region is the continuous system connected with the wells, whereas the second region only supports the first region with locally feeding fluid. These regions indicate fractures and matrix that have different properties of fluid storage and conductivity. This concept presented by Barrenblatt et al. (1960)

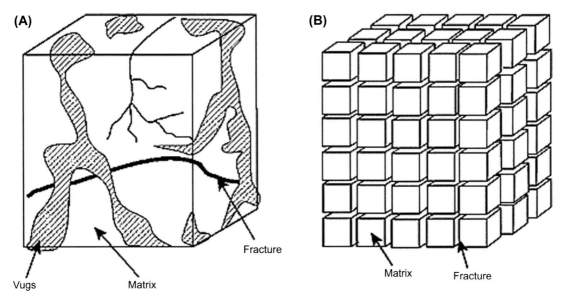

FIG. 2.11 **(A)** Actual reservoir system and **(B)** idealized orthogonal dual porosity model. (Credit: Warren, J. E., & Root, P. J. (1963). The behavior of naturally fractured reservoirs. *Society of Petroleum Engineers Journal*, 3(03), 245–255. https://doi.org/10.2118/426-PA.)

was based on heat conduction problem in a composite medium, which consists of a continuous, high-conductivity material and low-conductivity substance dispersed in the form of discrete particles (Stewart, 2011). The mechanism of transient heat conduction problem in this composite medium is mathematically similar to the pressure behavior in a dual porosity model.

A naturally fractured reservoir is comprised of complicated rock structure surrounded by irregular vug and natural fracture systems. To simplify this actual reservoir system, Warren and Root (1963) first presented the idealized system with an orthogonal set of cubic matrix blocks with intersecting fractures as shown in Fig. 2.11. In several studies, it has been proven that a simplified homogeneous dual porosity model can constitute to analyze an actual heterogeneous naturally fractured reservoir (Kazemi, Merrill, Porterfield, & Zeman, 1976; Lim & Aziz, 1995; Warren & Root, 1963). In addition, to make the dual porosity model more realistic, several improved models of dual porosity system are presented: the subdomain method (Gilman, 1986; Saidi, 1983), the multiple interacting continua (MINC) method (Pruess & Narasimhan, 1982), pseudocapillary pressure and relative permeability techniques (Dean & Lo, 1988; Thomas, Dixon, & Pierson, 1983), and dual permeability models

(Blaskovich, Cain, Sonier, Waldren, & Webb, 1983; Hill & Thomas, 1985; Dean & Lo, 1988). In the classic dual porosity system, reservoir fluids flow through the interconnected fracture networks, whereas matrix plays only a storage role and feeds fluids to the fractures. However, in realistic shale reservoir system, fluid flow can be performed in the matrix. Dual permeability model, which considers fluid flow from the matrix to matrix blocks as well as flows from the matrix to fracture and from fracture to fracture blocks. Detailed numerical models of dual porosity and dual permeability systems are presented in Section 4.1.

The dual porosity model assumes that the matrix has ample storage capacity but low permeability compared with the natural fracture system. The fractures are assumed to have little storage capacity but high permeability relative to the matrix system. Warren and Root (1963) introduced two dual porosity variables such as storativity ratio, ω, and interporosity flow coefficient, λ, which are used to describe the naturally fractured system. The storativity ratio, ω, is defined as:

$$\omega = \frac{\phi_{fb}c_f}{\phi_{fb}c_f + \phi_{mb}c_m}, \tag{2.1}$$

where ϕ is the porosity and c is the compressibility. Subscripts f and m indicate fracture and matrix. Subscript b presents bulk property. In the dual porosity system, properties based on total volume give representative values rather than intrinsic properties, which would be measured by direct measurements such as core experiments. The bulk fracture porosity and permeability can be calculated with the intrinsic properties given below:

$$\phi_{fb} = \frac{V_f}{V_{f+m}} \phi_{fi}, \tag{2.2}$$

$$k_{fb} = \frac{V_f}{V_{f+m}} k_{fi}, \tag{2.3}$$

where subscription i indicates intrinsic property. Consequently, the storativity ratio means the relative fracture storage capacity in the dual porosity reservoir. The interporosity flow coefficient, λ, indicates the dynamics of the fluid exchange between matrix and fractures. The interporosity flow coefficient is defined as follows:

$$\lambda = \alpha r_w^2 \frac{k_{mb}}{k_{fb}}, \tag{2.4}$$

where

$$\alpha = \frac{4n(n+2)}{h_m^2}, \tag{2.5}$$

α is the parameter characteristics of the system geometry, or shape factor, n the number of normal fracture planes, and h_m the thickness of the matrix block. The number of normal fracture planes, $n = 1, 2$, and 3 indicate the idealized slab, columns, and cube models, respectively. The interporosity flow coefficient is a parameter how smoothly fluid flows between matrix and fractures. This important dimensionless group, interporosity flow coefficient, dominates the start of the fluid flow from the matrix. This parameter takes into account both matrix geometry and the permeability ratio of matrix to fracture systems. As interporosity flow coefficient decreases, fluid flow from matrix to fractures will be delayed more. The properties of storativity ratio and interporosity flow coefficient are used for calculation of dual porosity responses in both pseudosteady state and transient flow models.

There are two common dual porosity models to describe pressure responses in the naturally fractures reservoirs: pseudosteady state and transient flow models. The pseudosteady state flow model was presented by Warren and Root (1963), and transient flow model was provided by de Swaan (1976) and Serra, Reynolds, and Raghavan (1983). Although matrix flow is almost transient, it may behave like pseudosteady state when there is an obstacle for fluid flow from the low-permeability matrix to high-permeability fractures. Therefore, pseudosteady state model is more common than transient model because of effective computational cost.

The main assumption of pseudosteady state flow model is that the matrix pressure decreases at the same rate in all points so that fluid flow from the matrix to the fracture is proportional to the difference between matrix and adjacent fracture pressure. In this model, unsteady state flow does not present, and pseudosteady state flow starts from the beginning of the flow. Fig. 2.12 shows the characteristic pressure response of pseudosteady state flow model in the semilog graph. In the semilog graph, pressure behavior is characterized by the first straight line, a transition that looks like a straight line of nearly a zero slope, and a final straight line displaying the same slope as the first. The first

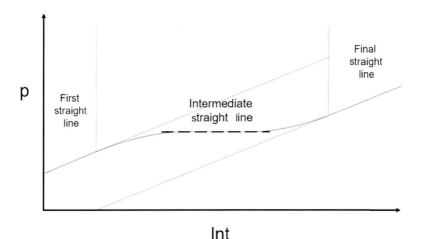

FIG. 2.12 Pressure response of pseudosteady state dual porosity flow model in the semilog plot.

straight line usually shows very short duration and represents the fracture system alone. The equations of this line are as follows:

$$p_{wD} = \frac{1}{2}(\ln t_D) - \ln \omega + \ln \frac{4}{\gamma} \quad \text{or} \quad (2.6)$$

$$p_{wf} = p_i - \frac{q_{sc}B\mu}{4\pi k_{fb}h}\left(\ln t + \ln \frac{4k_{fb}}{\phi_{fb}c_f \mu r_w^2 \gamma}\right), \quad (2.7)$$

where p_{wD} is the dimensionless well flowing pressure, t_D the dimensionless time, γ the exponential of Euler's constant, 1.781 or $e^{0.5772}$, p_{wf} the well flowing pressure p_i the initial reservoir pressure, q_{sc} the rate at the standard condition, B the formation volume factor, μ the viscosity, and r_w the well radius. In initial straight line stage, the reservoir behaves like a homogeneous formation because fluids flow only in fracture system without contribution from the matrix system. After a transitional stage that presents a discrete pressure drop, the fluid starts to flow from the matrix into the fracture, and it presents a nearly flat line in semilog plot. The final straight line represents total system behavior, and the equations of this period are given by

$$p_{wD} = \frac{1}{2}\left(\ln t_D + \ln \frac{4}{\gamma}\right) \quad \text{or} \quad (2.8)$$

$$p_{wf} = p_i - \frac{q_{sc}B\mu}{4\pi k_{fb}h}\left[\ln t + \ln \frac{4k_{fb}}{(\phi c_t)_{m+f}\mu r_w^2 \gamma}\right], \quad (2.9)$$

where c_t is the total compressibility. The separation between the lines is $ln\omega$, which becomes larger in absolute value as ω become smaller. In the final stage, matrix and fracture reach an equilibrium condition. At this time, the reservoir also behaves like a homogeneous system although this system is comprised of both matrix and fractures. Because the permeability of fractures is extremely greater than that of matrix, slope of the second straight line is almost same with that of the initial straight line.

Transient or unsteady state flow is more common than pseudosteady state flow in the matrix system. In the transient dual porosity system, pressure drawdown starts at the interface of matrix and fracture, and then it moves further into the center of the matrix as time goes on. In this case, pseudosteady state flow can be achieved only at a late time. Fig. 2.13 presents a characteristic pressure response of the transient flow model in semilog graph. The shape is different from that of pseudosteady state flow model. In this graph, three distinct flow regimes are identified. The first flow regime, which presents an initial straight line, occurs when fluids flow in only fractures. In the second flow regime, the fluid flow from the matrix into the fracture starts, and it continues until the fluid flow from the matrix to fracture reaches equilibrium. The straight line of the second flow regime in a semilog plot shows a slope of approximately one-half compared with that of first and last flow regimes (Serra et al., 1983). After fluid flow from the matrix to fractures reaches equilibrium, the last flow regime, which is dominant from the matrix through the wellbore, is presented.

Discrete Fracture Model

Although dual porosity models are commonly used for the simulation of naturally fractured shale reservoirs, it may not be adequate in some complex fracture networks. Because dual porosity systems simplify complex

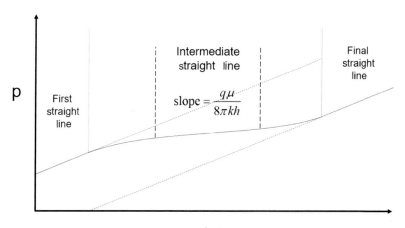

FIG. 2.13 Pressure response of transient dual porosity flow model in the semilog plot

connectivity and scale-dependent heterogeneity, they present a vast difference between reality and models. Discrete fracture models have been developed to overcome the inherent problems in dual porosity models (Hui, Karimi-Fard, Mallison, & Durlofsky, 2018; Karimi-Fard, Durlofsky, & Aziz, 2004; Moinfar, Narr, HuiMallison, & Lee, 2011). Discrete fracture models fully define the individual fracture geometry and simulate the fluid flow through the fracture geometry. Most of discrete fracture models use the unstructured grid system to present realistic fracture system geometry. Discrete fracture models are easily adaptable and updatable because discrete fracture models are not constrained by grid-defined fracture geometry (Moinfar et al., 2011). In addition, interpretation of fluid exchange between matrix and fracture system is more straightforward because of directly visible fracture geometry. Although discrete fracture models have fatal disadvantages such as difficulty and the high cost of data acquisition and numerical computation in a full-field reservoir model, they should be considered for accurate modeling of more complex fracture systems.

Numerous studies have been presented for discrete fracture model to simulate the flow behavior in naturally fractured reservoirs. The finite element method was used to simulate single-phase flow in fractured reservoirs by Noorishad and Mehran (1982) and Baca, Arnett, and Langford (1984). The finite element method is extended to two-phase flow in fractured reservoirs by Kim and Deo (2000) and Karimi-Fard and Firoozabadi (2003). The interest on discrete fracture models has grown particularly in recent years. Based on methods for fracture generation, they can be separated into two types: unstructured discrete fracture model and embedded discrete fracture model (Hui et al., 2018; Moinfar et al., 2011). In unstructured discrete fracture models, explicit features are considered, and the surrounding matrix rock is gridded to conform to the fractures (Fu, Yang, & Deo, 2005; Geiger-Boschung, Matthäi, Niessner, & Helmig, 2009; Hoteit & Firoozabadi, 2005; Karimi-Fard et al. 2004; Matthäi, Mezentsev, & Belayneh, 2005; Mallison, Hui, & Narr, 2010). In embedded discrete fracture models, fractures are embedded within matrix blocks, and matrix-fracture interactions are dominated by a single connection with transmissibility with suitable local flow assumptions (Fumagalli, Pasquale, Zonca, & Micheletti, 2016; Li & Lee, 2008; Moinfar, Varavei, Sepehrnoori, & Johns, 2014; Vitel & Souche, 2007).

Unstructured discrete fracture model is based on unstructured gridding with local refinement near fractures. In this model, the fracture geometry,

connectivity, and fracture-matrix network are generated explicitly without simplifying assumptions. It is necessary to apply an unstructured discretization scheme to model the complex fractured porous medium accurately. However, reservoir simulation using unstructured discrete fracture model presents considerably expensive calculation cost because of the high resolution of fracture and matrix system. Even if several papers demonstrated that it is possible to apply unstructured discrete fracture models in actual full-field applications using state-of-the-art linear solvers and parallelization (Hui, Glass, Harris, & Jia, 2013; Lim, Hui, & Mallison, 2009), unstructured discrete fracture models are still expensive owing to the quantification of uncertainty, such as various fracture generations, reservoir parameters, and production scenarios.

Hui et al. (2018) presented full-field simulation results considering a comprehensive set of unstructured discrete fracture method with efficiently simulating recovery processes in the naturally fractured reservoir. The method includes grid generation, discretization, and aggregation-based upscaling of discrete fracture models. The method of Mallison et al. (2010) was used for grid generation. This specialized grid generation technique provides high-quality polyhedral and polygonal cells while allowing precise control of the minimum cell size in the grid. This is achieved by allowing slight modifications in the geometry of fracture intersections that ensure high grid quality without introducing arbitrarily small cells. For discretization, treatment suggested by Karimi-Fard et al. (2004) and Karimi-Fard and Durlofsky (2016) was used. The discrete fracture model is discretized with a finite-volume method that applies an optimized two-point flux approximation. Coarse-grid control models are generated by aggregating fine-grid cells. General flow-based upscaling procedures calculate the coarse-scale transmissibilities. This aggregation-based upscaling idea has been adopted by Fumagalli, Zonca, and Formaggia (2017) and Lie, Møyner, and Natvig (2017). Hui et al. (2018) presented numerical results for water flood, sour-gas injection, and primary gas-condensate production for fracture models with matrix and fracture heterogeneities (Fig. 2.14). Although the accuracy of coarse-scale model decreases with increasing levels of coarsening, as expected, results demonstrated that two orders of magnitude of speedup can be achieved with approximately less than 10% of error with their methodology.

To reduce the challenges of grid generation from unstructured discrete fracture model, embedded discrete fracture model, which includes simpler structured

(A)   **(B)**

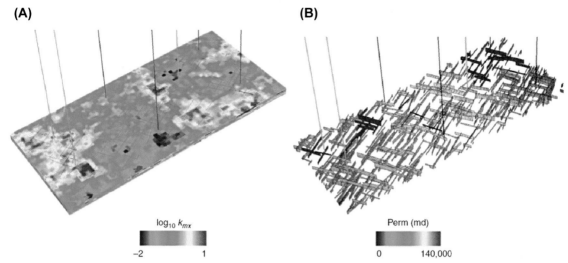

FIG. 2.14 Schematic view of discrete fracture model. **(A)** matrix permeability, **(B)** fracture permeability. (Credit: Hui, M.-H., Karimi-Fard, M., Mallison, B., & Durlofsky, L.J. (2018). A general modeling framework for simulating complex recovery processes in fractured reservoirs at different resolutions. *SPE Journal, 23*(02), 598–613. https://doi.org/10.2118/182621-PA.)

fractures in a conventional grid, is presented. Embedded discrete fracture model provides the accurate representation of fractures and keeps the matrix grid in uniformly refined and relatively coarse form. However, in this model, problems such as strong pressure and saturation gradients in the matrix would significantly affect the accuracy and efficiency of numerical computation so that some adjustive matrix grid refinement may be needed to describe detailed physics in reservoirs. Recently, Tene et al. (2017) presented the extension of embedded discrete fracture model. Representing matrix-fracture interactions more generally, it is similar to the unstructured discrete fracture model.

REFERENCES

Ahmed, U., & Meehan, D. N. (2016). *Unconventional oil and gas resources: Exploitation and development*. CRC Press.

Alharthy, N. S., Weldu Teklu, T. W., Nguyen, T. N., Kazemi, H., & Graves, R. M. (2016). Nanopore compositional modeling in unconventional shale reservoirs. *SPE Reservoir Evaluation & Engineering, 19*(03), 415–428. https://doi.org/10.2118/166306-PA.

Allaby, A., & Allaby, M. (1999). *A dictionary of Earth sciences*. Oxford University Press.

Ambrose, R. J., Hartman, R. C., Diaz Campos, M., Akkutlu, I. Y., & Sondergeld, C. (2010). New pore-scale considerations for shale gas in place calculations. In *Paper presented at the SPE unconventional gas conference, Pittsburgh, Pennsylvania, USA*. https://doi.org/10.2118/131772-MS.

Anderson, B., Barber, T., Lüling, M., Sen, P., Taherian, R., & Klein, J. (2008). Identifying potential gas-producing shales from large dielectric permittivities measured by induction quadrature signals. In *Paper presented at the 49th annual logging symposium, Austin, Texas*.

Archie, G. E. (1942). The electrical resistivity log as an aid in determining some reservoir characteristics. *Transactions of the AIME, 146*(01), 54–62. https://doi.org/10.2118/942054-G.

Baca, R. G., Arnett, R. C., & Langford, D. W. (1984). Modeling fluid flow in fractured-porous rock masses by finite element techniques. *International Journal for Numerical Methods in Fluids, 4*.

Ballard, B. D. (2007). Quantitative mineralogy of reservoir rocks using fourier transform infrared spectroscopy. In *Paper presented at the SPE annual technical conference and exhibition, Anaheim, California, USA*. https://doi.org/10.2118/113023-STU.

Barenblatt, G. I., Zheltov, I. P., & Kochina, I. N. (1960). Basic concepts in the theory of seepage of homogeneous liquids in fissured rocks [strata]. *Journal of Applied Mathematics and Mechanics, 24*(5), 1286–1303. https://doi.org/10.1016/0021-8928(60)90107-6.

Barsotti, E., Tan, S. P., Saraji, S., Piri, M., & Chen, J.-H. (2016). A review on capillary condensation in nanoporous media: Implications for hydrocarbon recovery from tight reservoirs. *Fuel, 184*, 344–361. https://doi.org/10.1016/j.fuel.2016.06.123.

Bates, R. L., Jackson, J. A., & Institute, A. G. (1984). *Dictionary of geological terms*. Anchor Press/Doubleday.

Bernard, S., Horsfield, B., Schulz, H.-M., Wirth, R., Schreiber, A., & Sherwood, N. (2012a). Geochemical evolution of organic-rich shales with increasing maturity: A STXM and TEM study of the posidonia shale (lower Toarcian, Northern Germany). *Marine and Petroleum Geology, 31*(1), 70−89. https://doi.org/10.1016/j.marpetgeo.2011.05.010.

Bernard, S., Wirth, R., Schreiber, A., Schulz, H.-M., & Horsfield, B. (2012b). Formation of nanoporous pyrobitumen residues during maturation of the Barnett shale (Fort Worth Basin). *International Journal of Coal Geology, 103*, 3−11. https://doi.org/10.1016/j.coal.2012.04.010.

Bhuyan, K., & Passey, Q. R. (1994). Clay estimation from Gr and neutron-density porosity logs. In *Paper presented at the SPWLA 35th annual logging symposium, Tulsa, Oklahoma.*

Blaskovich, F. T., Cain, G. M., Sonier, F., Waldren, D., & Webb, S. J. (1983). A multicomponent isothermal system for efficient reservoir simulation. In *Paper presented at the Middle East oil technical conference and exhibition, Manama, Bahrain.* https://doi.org/10.2118/11480-MS.

Bohacs, K. M. (1998). Contrasting expressions of depositional sequences in mudstones from marine to non-marine environs. In *Shales and mudstones*. Stuttgart: E. Schweizerbart.

Bohacs, K. M., & Miskell-Gerhardt, K. (1998). Well-log expression of lake strata; controls of lake-basin type and provenance, contrasts with marine strata. In *AAPG annual convention and exhibition, Salt Lake City, Utah, USA.*

Bowker, K. (2003). *Recent development of the Barnett Shale play* (Vol. 42). Fort Worth Basin.

Bust, V. K., Majid, A., Oletu, J. U., & Worthington, P. F. (2011). The petrophysics of shale gas reservoirs: Technical challenges and pragmatic solutions. In *Paper presented at the international petroleum technology conference, Bangkok, Thailand.* https://doi.org/10.2523/IPTC-14631-MS.

Chen, J.-H., Mehmani, A., Li, B., Georgi, D., & Jin, G. (2013). Estimation of total hydrocarbon in the presence of capillary condensation for unconventional shale reservoirs. In *Paper presented at the SPE Middle East oil and gas show and conference, Manama, Bahrain.* https://doi.org/10.2118/164468-MS.

Choquette, P. W., & Pray, L. C. (1970). Geologic nomenclature and classification of porosity in sedimentary carbonates. *Aapg Bulletin, 54*(2), 44.

Curtis, M. E., Ambrose, R. J., & Sondergeld, C. H. (2010). Structural characterization of gas shales on the micro- and nanoscales. In *Paper presented at the Canadian unconventional resources and international petroleum conference, Calgary, Alberta, Canada.* https://doi.org/10.2118/137693-MS.

Dayal, A., & Mani, D. (2017). *Shale gas: Exploration and environmental and economic impacts.*

de Swaan, O.,A. (1976). Analytic solutions for determining naturally fractured reservoir properties by well testing. *Society of Petroleum Engineers Journal, 16*(03), 117−122. https://doi.org/10.2118/5346-PA.

Dean, R. H., & Lo, L. L. (1988). Simulations of naturally fractured reservoirs. *SPE Reservoir Engineering, 3*(02), 638−648. https://doi.org/10.2118/14110-PA.

Desbois, G., Urai, J., & Kukla, P. (2009). Morphology of the pore space in claystones-evidence from BIB/FIB ion beam sectioning and cryo-SEM observations. *eEarth, 4*, 15−22. https://doi.org/10.5194/ee-4-15-2009.

Dresser, Atlas. (1979). *Dresser atlas log interpretation charts.* Houston, Tex: Dresser Industries.

EIA. (2016). *Maps: Oil and gas exploration, resources, and production.* https://www.eia.gov/maps/maps.htm.

Fertl, W. H., & Rieke, H. H., III (1980). Gamma ray spectral evaluation techniques identify fractured shale reservoirs and source-rock characteristics. *Journal of Petroleum Technology, 32*(11), 2053−2062. https://doi.org/10.2118/8454-PA.

Fishman, N., Guthrie, J., & Honarpour, M. (2013). The stratigraphic distribution of hydrocarbon storage and its effect on producible hydrocarbons in the Eagle Ford formation, south Texas. In *Paper presented at the unconventional resources technology conference, Denver, Colorado, USA.*

Fishman, N. S., Hackley, P. C., Lowers, H. A., Hill, R. J., Egenhoff, S. O., Eberl, D. D., et al. (2012). The nature of porosity in organic-rich mudstones of the upper Jurassic Kimmeridge clay formation, north sea, offshore United Kingdom. *International Journal of Coal Geology, 103*, 32−50. https://doi.org/10.1016/j.coal.2012.07.012.

Fu, Y., Yang, Y. K., & Deo, M. (2005). Three-dimensional, three-phase discrete-fracture reservoir simulator based on control volume finite element (CVFE) formulation. In *Paper presented at the SPE reservoir simulation symposium, The Woodlands, Texas.* https://doi.org/10.2118/93292-MS.

Fumagalli, A., Pasquale, L., Zonca, S., & Micheletti, S. (2016). An upscaling procedure for fractured reservoirs with embedded grids. *Water Resources Research, 52*(8), 6506−6525. https://doi.org/10.1002/2015WR017729.

Fumagalli, A., Zonca, S., & Formaggia, L. (2017). Advances in computation of local problems for a flow-based upscaling in fractured reservoirs. *Mathematics and Computers in Simulation, 137*, 299−324. https://doi.org/10.1016/j.matcom.2017.01.007.

Gale, J. F. W., Laubach, S. E., Olson, J. E., Eichhubl, P., & Fall, A. (2014). Natural fractures in shale: A review and new observationsNatural fractures in shale: A review and new observations. *AAPG Bulletin, 98*(11), 2165−2216. https://doi.org/10.1306/08121413151.

Geiger-Boschung, S., Matthäi, S. K., Niessner, J., & Helmig, R. (2009). Black-oil simulations for three-component, three-phase flow in fractured porous media. *SPE Journal, 14*(02), 338−354. https://doi.org/10.2118/107485-PA.

Gilman, J. R. (1986). An efficient finite-difference method for simulating phase segregation in the matrix blocks in double-porosity reservoirs. *SPE Reservoir Engineering, 1*(04), 403−413. https://doi.org/10.2118/12271-PA.

Givens, N., & Zhao, H. (2004). The Barnett shale: Not so simple after all. In *AAPG Annual Meeting, Dallas, Texas.*

Harper, J. A. (1999). Pennsylvania geological survey and pittsburgh geological society. In C. H. Shultz (Ed.), *The geology of Pennsylvania* (pp. 108−127). Pennsylvania Geological Survey; Pittsburgh Geological Society.

He, J., Ding, W., Zhang, J., Li, A., Zhao, W., & Peng, D. (2016). Logging identification and characteristic analysis of marine−continental transitional organic-rich shale in the Carboniferous-Permian strata, Bohai Bay Basin. *Marine and Petroleum Geology, 70,* 273−293. https://doi.org/10.1016/j.marpetgeo.2015.12.006.

Hill, A. C., & Thomas, G. W. (1985). A new approach for simulating complex fractured reservoirs. In *Paper presented at the Middle East oil technical conference and exhibition, Bahrain.* https://doi.org/10.2118/13537-MS.

Hoteit, H., & Firoozabadi, A. (2005). Multicomponent fluid flow by discontinuous Galerkin and mixed methods in unfractured and fractured media. *Water Resources Research, 41*(11). https://doi.org/10.1029/2005WR004339.

Houben, M. E., Desbois, G., & Urai, J. L. (2014). A comparative study of representative 2D microstructures in Shaly and Sandy facies of Opalinus Clay (Mont Terri, Switzerland) inferred form BIB-SEM and MIP methods. *Marine and Petroleum Geology, 49,* 143−161. https://doi.org/10.1016/j.marpetgeo.2013.10.009.

Hui, M.-H., Glass, J., Harris, D., & Jia, C. (2013). Discrete natural fracture uncertainty modelling for produced water mitigation: Chuandongbei gas project, sichuan, China. In *Paper presented at the international petroleum technology conference, Beijing, China.* https://doi.org/10.2523/IPTC-16814-MS.

Hui, M.-H., Karimi-Fard, M., Mallison, B., & Durlofsky, L. J. (2018). A general modeling framework for simulating complex recovery processes in fractured reservoirs at different resolutions. *SPE Journal, 23*(02), 598−613. https://doi.org/10.2118/182621-PA.

Hunt, J. M. (1996). *Petroleum geochemistry and geology.* W.H. Freeman.

Hyne, N. J. (1991). *Dictionary of petroleum exploration, drilling & production.* PennWell Publishing Company.

Jacobi, D. J., Gladkikh, M., LeCompte, B., Hursan, G., Mendez, F., Longo, J., et al. (2008). Integrated petrophysical evaluation of shale gas reservoirs. In *Paper presented at the CIPC/SPE gas technology symposium 2008 joint conference, Calgary, Alberta, Canada.* https://doi.org/10.2118/114925-MS.

Jarvie, D. M. (1991). Total Organic Carbon (TOC) analysis: In: Source and migration processes and evaluation techniques. *APG Bulletin,* 113−118.

Jarvie, D. M., Hill, R. J., Ruble, T. E., & Pollastro, R. M. (2007). Unconventional shale-gas systems: The Mississippian Barnett Shale of north-central Texas as one model for thermogenic shale-gas assessment. *American Association of Petroleum Geologists Bulletin, 91*(4), 475−499. https://doi.org/10.1306/12190606068.

Jiang, Z., Zhang, W., Liang, C., Wang, Y., Liu, H., & Chen, X. (2016). Basic characteristics and evaluation of shale oil reservoirs. *Petroleum Research, 1*(2), 149−163. https://doi.org/10.1016/S2096-2495(17)30039-X.

Karimi-Fard, M., & Durlofsky, L. J. (2016). A general gridding, discretization, and coarsening methodology for modeling flow in porous formations with discrete geological features. *Advances in Water Resources, 96,* 354−372. https://doi.org/10.1016/j.advwatres.2016.07.019.

Karimi-Fard, M., Durlofsky, L. J., & Aziz, K. (2004). An efficient discrete-fracture model applicable for general-purpose reservoir simulators. *SPE Journal, 9*(02), 227−236. https://doi.org/10.2118/88812-PA.

Karimi-Fard, M., & Firoozabadi, A. (2003). Numerical simulation of water injection in fractured media using the discrete-fracture model and the Galerkin method. *SPE Reservoir Evaluation & Engineering, 6*(02), 117−126. https://doi.org/10.2118/83633-PA.

Kazemi, H., Merrill, L. S., Jr., Porterfield, K. L., & Zeman, P. R. (1976). Numerical simulation of water-oil flow in naturally fractured reservoirs. *Society of Petroleum Engineers Journal, 16*(06), 317−326. https://doi.org/10.2118/5719-PA.

Kim, J.-G., & Deo, M. D. (2000). Finite element, discrete-fracture model for multiphase flow in porous media. *AIChE Journal, 46*(6), 1120−1130. https://doi.org/10.1002/aic.690460604.

Kochenov, A. V., & Baturin, G. N. (2002). The paragenesis of organic matter, phosphorus, and uranium in marine sediments. *Lithology and Mineral Resources, 37*(2), 107−120. https://doi.org/10.1023/A:1014816315203.

Ko, L. T., Loucks, R. G., Ruppel, S. C., Zhang, T., & Peng, S. (2017). Origin and characterization of Eagle Ford pore networks in the south Texas Upper Cretaceous shelf. *AAPG Bulletin, 101*(3), 387−418. https://doi.org/10.1306/08051616035.

Ko, L. T., Zhang, T., Loucks, R. G., Ruppel, S. C., & Shao, D. (2015). *Pore evolution in the Barnett, Eagled Ford (Boquillas), and Woodford mudrocks based on gold-tube pyrolysis thermal maturation.* Denver, Colorado, USA.

Kuila, U., & Prasad, M. (2011). Understanding pore-structure and permeability in shales. In *Paper presented at the SPE annual technical conference and exhibition, Denver, Colorado, USA.* https://doi.org/10.2118/146869-MS.

Kuila, U., & Prasad, M. (2013). Specific surface area and pore-size distribution in clays and shales. *Geophysical Prospecting, 61*(2), 341−362. https://doi.org/10.1111/1365-2478.12028.

Larese, R. E., & Heald, M. T. (1976). *Petrography of selected Devonian shale core samples from the CGTC No. 20403 and CGSC No. 11940 wells.* West Virginia. United States: Lincoln and Jackson Counties.

Lie, K.-A., Møyner, O., & Natvig, J. R. (2017). Use of multiple multiscale operators to accelerate simulation of complex geomodels. *SPE Journal, 22*(06), 1929−1945. https://doi.org/10.2118/182701-PA.

Li, L., & Lee, S. H. (2008). Efficient field-scale simulation of black oil in a naturally fractured reservoir through discrete fracture networks and homogenized media. *SPE Reservoir Evaluation & Engineering, 11*(04), 750−758. https://doi.org/10.2118/103901-PA.

Lim, K. T., & Aziz, K. (1995). Matrix-fracture transfer shape factors for dual-porosity simulators. *Journal of Petroleum Science and Engineering, 13*(3), 169−178. https://doi.org/10.1016/0920-4105(95)00010-F.

Lim, K.-T., Hui, M.-H., & Mallison, B. T. (2009). A next-generation reservoir simulator as an enabling technology for a complex discrete fracture modeling workflow. In *Paper presented at the SPE annual technical conference and exhibition, New Orleans, Louisiana.* https://doi.org/10.2118/124980-MS.

Loucks, R., & Reed, R. (2014). Scanning-electron-microscope petrographic evidence for distinguishing organic matter pores associated with depositional organic matter versus migrated organic matter in mudrocks. *GCAGS Journal, 3,* 51−60.

Loucks, R. G., Reed, R., Ruppel, S. C., & Hammes, U. (2012). Spectrum of pore types and networks in mudrocks and a descriptive classification for matrix-related mudrock pores. *AAPG Bulletin, 96*(6), 1071−1098.

Loucks, R. G., Reed, R. M., Ruppel, S. C., & Jarvie, D. M. (2009). Morphology, genesis, and distribution of nanometer-scale pores in diliceous mudstones of the Mississippian Barnett Shale. *Journal of Sedimentary Research, 79*(12), 848−861.

Loucks, R., & Ruppel, S. (2007). *Mississippian Barnett Shale: Lithofacies and depositional setting of a deep-water shale-gas succession in the Fort Worth Basin* (Vol. 91). Texas.

Lu, S., Huang, W., Chen, F., Li, J., Wang, M., Xue, H., et al. (2012). Classification and evaluation criteria of shale oil and gas resources: Discussion and application. *Petroleum Exploration and Development, 39*(2), 268−276. https://doi.org/10.1016/S1876-3804(12)60042-1.

Macquaker, J. H. S., Keller, M. A., & Davies, S. J. (2010). Algal blooms and Marine snow: Mechanisms that enhance preservation of organic carbon in ancient fine-grained sediments. *Journal of Sedimentary Research, 80*(11), 934−942. https://doi.org/10.2110/jsr.2010.085.

Mallison, B., Hui, M. H., & Narr, W. (2010). *Practical gridding algorithms for discrete fracture modeling workflows.*

Mani, D., Patil, D. J., & Dayal, A. M. (2015). Organic properties and hydrocarbon generation potential of shales from few sedimentary basins of India. In S. Mukherjee (Ed.), *Petroleum Geosciences: Indian Contexts* (pp. 99−126). Cham: Springer International Publishing.

Matthäi, S. K., Mezentsev, A., & Belayneh, M. (2005). Control-volume finite-element two-phase flow experiments with fractured rock represented by unstructured 3D hybrid meshes. In *Paper presented at the SPE reservoir simulation symposium, The Woodlands, Texas.* https://doi.org/10.2118/93341-MS.

Milici, R. C., & Swezey, C. S. (2006). Assessment of Appalachian Basin oil and gas resources:devonian shale - Middle and upper paleozoic total petroleum system. In *Open-file report.*

Milliken, K. L., Ko, L. T., Pommer, M. E., & Marsaglia, K. M. (2014). SEM petrography of Eastern Mediterranean sapropels: Analogue data for assessing organic matter in oil and gas shales. *Journal of Sedimentary Research, 84*(11), 961−974.

Milliken, K. L., & Reed, R. M. (2010). Multiple causes of diagenetic fabric anisotropy in weakly consolidated mud,

Nankai accretionary prism, IODP Expedition 316. *Journal of Structural Geology, 32*(12), 1887−1898. https://doi.org/10.1016/j.jsg.2010.03.008.

Moinfar, A., Narr, W., Hui, M.-H., Mallison, B. T., & Lee, S. H. (2011). Comparison of discrete-fracture and dual-permeability models for multiphase flow in naturally fractured reservoirs. In *Paper presented at the SPE reservoir simulation symposium, The Woodlands, Texas, USA.* https://doi.org/10.2118/142295-MS.

Moinfar, A., Varavei, A., Sepehrnoori, K., & Johns, R. T. (2014). Development of an efficient embedded discrete fracture model for 3D compositional reservoir simulation in fractured reservoirs. *SPE Journal, 19*(02), 289−303. https://doi.org/10.2118/154246-PA.

Mondol, N. H., Bjørlykke, K., Jahren, J., & Høeg, K. (2007). Experimental mechanical compaction of clay mineral aggregates—changes in physical properties of mudstones during burial. *Marine and Petroleum Geology, 24*(5), 289−311. https://doi.org/10.1016/j.marpetgeo.2007.03.006.

Montgomery, S. L., Jarvie, D. M., Bowker, K. A., & Pollastro, R. M. (2005). Mississippian Barnett Shale, Fort Worth basin, north-central Texas: Gas-shale play with multi-trillion cubic foot potential. *American Association of Petroleum Geologists Bulletin, 89*(2), 155−175. https://doi.org/10.1306/09170404042.

Morley, C. K., von Hagke, C., Hansberry, R., Collins, A., Kanitpanyacharoen, W., & King, R. (2018). Review of major shale-dominated detachment and thrust characteristics in the diagenetic zone: Part II, rock mechanics and microscopic scale. *Earth-science Reviews, 176,* 19−50. https://doi.org/10.1016/j.earscirev.2017.09.015.

Myrow, P. M., & Southard, J. B. (1996). Tempestite deposition. *Journal of Sedimentary Research, 66*(5), 875−887. https://doi.org/10.1306/D426842D-2B26-11D7-8648000102C1865D.

Narr, W., Schechter, D. S., & Thompson, L. B. (2006). *Naturally fractured reservoir characterization.* Society of Petroleum Engineers.

NETL. (2011). *A comparative study of the Mississippian Barnett shale, Fort Worth Basin, and devonian Marcellus shale, Appalachian Basin.* U.S. Department of Energy.

Noorishad, J., & Mehran, M. (1982). An upstream finite element method for solution of transient transport equation in fractured porous media. *Water Resources Research, 18*(3), 588−596. https://doi.org/10.1029/WR018i003p00588.

Nuhfer, E. B., Vinopal, R. J., & Klanderman, D. S. (1979). *X-radiograph Atlas of lithotypes and other structures in the devonian shale sequence of West Virginia and Virginia.* Department of Energy, Morgantown Energy Technology Center.

Papazis, P. K. (2005). *Petrographic characterization of the Barnett shale.* Texas: Fort Worth Basin. http://worldcat.org.

Passey, Q. R., Bohacs, K., Esch, W. L., Klimentidis, R., & Sinha, S. (2010). From oil-prone source rock to gas-producing shale reservoir - geologic and petrophysical characterization of unconventional shale gas reservoirs. In *Paper presented at the international oil and gas conference and exhibition in China, Beijing, China.* https://doi.org/10.2118/131350-MS.

Passey, Q. R., Moretti, F. J., Kulla, J. B., Creaney, S., & Stroud, J. D. (1990). *A practical model for organic richness from porosity and resistivity logs* (Vol. 74).

Pemper, R. R., Han, X., Mendez, F. E., Jacobi, D., LeCompte, B., Bratovich, M., et al. (2009). The direct measurement of carbon in wells containing oil and natural gas using a pulsed neutron mineralogy tool. In *Paper presented at the SPE annual technical conference and Exhibition, New Orleans, Louisiana*. https://doi.org/10.2118/124234-MS.

Pemper, R. R., Sommer, A., Guo, P., Jacobi, D., Longo, J., Bliven, S., et al. (2006). A new pulsed neutron sonde for derivation of formation lithology and mineralogy. In *Paper presented at the SPE annual technical conference and exhibition, San Antonio, Texas, USA*. https://doi.org/10.2118/102770-MS.

Peters, K. E., & Cassa, M. R. (1994). Applied source rock geochemistry. In *AAPG memoir* (Vol. 60, pp. 93−120).

Plint, A. G., Macquaker, J. H. S., & Varban, B. (2012). Bedload transport of mud across a wide, storm-influenced ramp: Cenomanian-Turonian Kaskapau formation, Western Canada foreland basin–reply. *Journal of Sedimentary Research, 82*.

Pommer, M. E. (2014). Quantitative assessment of pore types and pore size distribution across thermal maturity, Eagle Ford Formation, South Texas. In *Master of science in geological sciences*. The University of Texas at Austin.

Pommer, M., & Milliken, K. (2015). Pore types and pore-size distributions across thermal maturity, Eagle Ford formation, southern Texas Pores across thermal maturity, Eagle Ford. *AAPG Bulletin, 99*(9), 1713−1744. https://doi.org/10.1306/03051514151.

Potter, P. E., Maynard, J. B., & Pryor, W. A. (1980). *Final report of special geological, geochemical, and petrological studies of the Devonian shales in the Appalachian Basin. Cincinnati University*. OH (USA): H.N. Fisk Laboratory of Sedimentology.

Pruess, K., & Narasimhan, T. N. (1982). *Practical method for modeling fluid and heat flow in fractured porous media, conference: 6. Symposium on resrevior simulation, New Orleans, LA, USA, 1 Feb 1982; other information: Portions of document are illegible*. CA (USA): Lawrence Berkeley Lab.

Quirein, J., Witkowsky, J., Truax, J. A., Galford, J. E., Spain, D. R., & Odumosu, T. (2010). Integrating core data and wireline geochemical data for formation evaluation and characterization of shale-gas reservoirs. In *Paper presented at the SPE annual technical conference and exhibition, Florence, Italy*. https://doi.org/10.2118/134559-MS.

Ramirez, T. R., Klein, J. D., Ron, B., & Howard, J. J. (2011). Comparative study of formation evaluation methods for unconventional shale gas reservoirs: Application to the Haynesville shale (Texas). In *Paper presented at the North American unconventional gas conference and exhibition, The Woodlands, Texas, USA*. https://doi.org/10.2118/144062-MS.

Reed, R., & Ruppel, S. (2012). Pore morphology and distribution in the cretaceous Eagle Ford shale, south Texas, USA. *Gulf Coast Association of Geological Societies, 62*, 599−603.

Rezaee, R. (2015). *Fundamentals of gas shale reservoirs*. Wiley.

Rickman, R., Mullen, M. J., Petre, J. E., Grieser, W. V., & Kundert, D. (2008). A practical use of shale petrophysics for stimulation design optimization: All shale plays are not clones of the Barnett shale. In *Paper presented at the SPE annual technical conference and exhibition, Denver, Colorado, USA*. https://doi.org/10.2118/115258-MS.

Rider, M. H. (2002). *The geological interpretation of well logs*. Rider-French Consulting.

Roen, J. B. (1993). Petroleum geology of the Devonian and Mississippian black shale of eastern North America. In R. C. Kepferle (Ed.), *Estimates of unconventional natural gas resources of the Devonian Shales of the appalachian basin*.

Rokosh, C. D., Pawlowicz, J. G., Berhane, H., Anderson, S. D. A., & Beaton, A. P. (2008). *What is shale gas? An introduction to shale-gas geology in Alberta*. Energy Resources Conservation Board.

Ross, D. J. K., & Bustin, R. M. (2006). Sediment geochemistry of the lower Jurassic Gordondale member, northeastern British columbia. *Bulletin of Canadian Petroleum Geology, 54*(4), 337−365. https://doi.org/10.2113/gscpgbull.54.4.337.

Rouquerol, J., Avnir, D., Fairbridge, C. W., Everett, D. H., Haynes, J. M., Pernicone, N., et al. (1994). Recommendations for the characterization of porous solids (Technical Report). In *Pure and Applied Chemistry*.

Ruessink, B. H., & Harville, D. G. (1992). Quantitative analysis of bulk mineralogy: The applicability and performance of XRD and FTIR. In *Paper presented at the SPE formation damage control symposium, Lafayette, Louisiana*. https://doi.org/10.2118/23828-MS.

Rutter, E., Mecklenburgh, J., & Taylor, K. (2017). Geomechanical and petrophysical properties of mudrocks: Introduction. *Geological Society, London, Special Publications, 454*(1), 1−13. https://doi.org/10.1144/sp454.16.

Saidi, A. M. (1983). Simulation of naturally fractured reservoirs. In *Paper presented at the SPE reservoir simulation symposium, San Francisco, California*. https://doi.org/10.2118/12270-MS.

Schieber, J. (2011). Reverse engineering mother nature — shale sedimentology from an experimental perspective. *Sedimentary Geology, 238*(1), 1−22. https://doi.org/10.1016/j.sedgeo.2011.04.002.

Schieber, J., Southard, J., & Kevin, T. (2007). Accretion of mudstone beds from migrating floccule ripples. *Science, 318*(5857), 1760−1763. https://doi.org/10.1126/science.1147001.

Serra, K., Reynolds, A. C., & Raghavan, R. (1983). New pressure transient analysis methods for naturally fractured reservoirs (includes associated papers 12940 and 13014). *Journal of Petroleum Technology, 35*(12), 2271−2283. https://doi.org/10.2118/10780-PA.

Sondergeld, C. H., Newsham, K. E., Comisky, J. T., Rice, M. C., & Rai, C. S. (2010). Petrophysical considerations in evaluating and producing shale gas resources. In *Paper presented at the SPE unconventional gas conference, Pittsburgh, Pennsylvania, USA*. https://doi.org/10.2118/131768-MS.

Sondergeld, C. H., & Rai, C. S. (1993). A new exploration tool: Quantitative core characterization. *Pure and Applied Geophysics, 141*(2), 249−268. https://doi.org/10.1007/BF00998331.

Sone, H., & Zoback, M. D. (2013). Mechanical properties of shale-gas reservoir rocks — Part 1: Static and dynamic elastic properties and anisotropy. *Geophysics, 78*(5), D381−D392. https://doi.org/10.1190/geo2013-0050.1.

Stewart, G. (2011). *Well test design & analysis.* PennWell.

Stow, D. A. V., Huc, A. Y., & Bertrand, P. (2001). Depositional processes of black shales in deep water. *Marine and Petroleum Geology, 18*(4), 491−498. https://doi.org/10.1016/S0264-8172(01)00012-5.

Tene, M., Bosma, S. B. M., Al Kobaisi, M. S., & Hajibeygi, H. (2017). Projection-based embedded discrete fracture model (pEDFM). *Advances in Water Resources, 105,* 205−216. https://doi.org/10.1016/j.advwatres.2017.05.009.

Thomas, L. K., Dixon, T. N., & Pierson, R. G. (1983). Fractured reservoir simulation. *Society of Petroleum Engineers Journal, 23*(01), 42−54. https://doi.org/10.2118/9305-PA.

Tissot, B. P., & Welte, D. H. (1984). *Petroleum formation and occurrence.* Springer Berlin Heidelberg.

Trabucho-Alexandre, J., Hay, W., & De Boer, P. (2012). *Phanerozoic environments of black shale deposition and the Wilson Cycle* (Vol. 3).

Velde, B. (1996). Compaction trends of clay-rich deep sea sediments. *Marine Geology, 133*(3), 193−201. https://doi.org/10.1016/0025-3227(96)00020-5.

Vitel, S., & Souche, L. (2007). Unstructured upgridding and transmissibility upscaling for preferential flow paths in 3D fractured reservoirs. In *Paper presented at the SPE reservoir simulation symposium, Houston, Texas, USA.* https://doi.org/10.2118/106483-MS.

Wang, G., & Carr, T. R. (2013). Organic-rich Marcellus shale lithofacies modeling and distribution pattern analysis in the Appalachian Basin Organic-rich shale lithofacies modeling, Appalachian Basin. *AAPG Bulletin, 97*(12), 2173−2205. https://doi.org/10.1306/05141312135.

Warren, J. E., & Root, P. J. (1963). The behavior of naturally fractured reservoirs. *Society of Petroleum Engineers Journal, 3*(03), 245−255. https://doi.org/10.2118/426-PA.

Wrightstone, G. (2009). Marcellus shale - geologic controls on production. In *AAPG annual convention.*

Young, R. N., & Southard, J. B. (1978). Erosion of fine-grained marine sediments: Sea-floor and laboratory experiments. *GSA Bulletin, 89*(5), 663−672. https://doi.org/10.1130/0016-7606(1978)89<663:EOFMSS>2.0.CO;2.

CHAPTER 3

Specific Mechanisms in Shale Reservoirs

ABSTRACT

Fluid flow in shale reservoirs is significantly different from that in conventional reservoirs because of specific features of shale such as ultralow permeability matrix, nanoscale pores, natural and induced fracture system, gas adsorption/desorption, and high sensitivity on geomechanic deformation. Although several kinds of literature presented various mechanisms for transport in shale reservoirs, it is still lack of understanding, and there is no integrative consensus of transport mechanisms. Especially, non-Darcy flow in hydraulic fractures, gas adsorption/desorption, microscale flow in nanopores, molecular diffusion, geomechanics, and confinement effects have attracted attention in these days. In this chapter, therefore, numerous specific transport mechanisms in shale reservoirs are introduced. Details of fluid transport mechanisms in shale reservoirs are presented from basic theory to various experimental and analytic correlations. Understanding of state-of-the-art approaches for transport in shales is of importance to improve hydrocarbon production in shale reservoirs.

NON-DARCY FLOW

In 1856, French engineer Henry Darcy published a detailed account of his work in improving the waterworks in Dijon. Among his works, the item of present interest pertains to a problem in designing a suitable filter for the system (Hubbert, 1956). Darcy needed to know how large a filter would be required for a given quantity of water per day so that he proceeded a series of experiments on the design of a filter. Consider an experimental apparatus shown in Fig. 3.1. A circular cylinder of cross-sectional area A is filled with sand and outfitted with a pair of manometers. During steady state, the inflow rate Q is equal to the outflow rate. The experiment shows that specific discharge, $v = \frac{Q}{A}$, is directly proportional to $\Delta h = h_1 - h_2$ and inversely proportional to l.

Darcy established that the following equation gives the rate of flow:

$$v = -K \frac{(h_2 - h_1)}{l},\qquad(3.1)$$

where v is superficial velocity, K a factor of proportionality or hydraulic conductivity, h_1 and h_2 the hydraulic head defined by height above a reference level of the water in manometer terminated above and below the sand, and l the thickness of the sand.

When the same experiments are carried out with various fluids of density, ρ, and viscosity, μ, for ideal porous media consisting of uniform grains of diameter, d, under a constant hydraulic gradient $\frac{dh}{dl}$, the following relationships are obtained:

$$v \propto d^2,\qquad(3.2)$$

$$v \propto \rho g,\qquad(3.3)$$

$$v \propto \frac{1}{\mu}.\qquad(3.4)$$

Combined with the original Darcy's law, these proportionalities lead to a new form of Darcy's law.

$$v = -\frac{Cd^2 \rho g}{\mu} \frac{dh}{dl}.\qquad(3.5)$$

The generalized form of Darcy's law used in petroleum engineering is shown below:

$$-\frac{dp}{dx} = \frac{\mu v}{k},\qquad(3.6)$$

where k is the permeability defined by $k = Cd^2$ and related to hydraulic conductivity by $K = \frac{k\rho g}{\mu}$. A permeability of one Darcy results in a flow of 1 cm/sec for a fluid of 1 centipoise under a gradient of 1 atmosphere per centimeter. In most cases of production in reservoirs, fluid flow can be described by Darcy's law.

Darcy's law cannot be used when the flow rate is high. The upper limit of Darcy's law is usually identified with Reynolds number, a dimensionless number expressing the ratio of inertial to viscous forces. The

Transport in ... Reservoirs https://doi.org/10.1016/B978-0-12-817860-7.00003-6

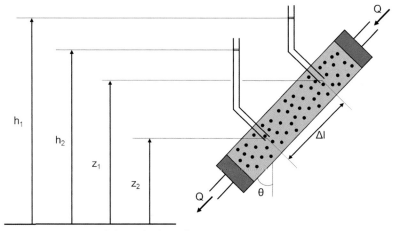

FIG. 3.1 Schematic view of Darcy's experimental apparatus.

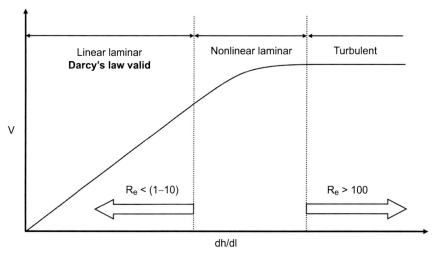

FIG. 3.2 Validity range of Darcy's law.

Reynolds number for the flow through porous media is defined as:

$$N_{Re} = \frac{\rho v d}{\mu}. \qquad (3.7)$$

The concept of a constant Darcy permeability, suggesting a linear relationship between flow rate and potential gradient, is valid when Reynolds number does not exceed some value between 1 and 10 (Fig. 3.2).

Empirical observations of the flow rate at higher differential pressures confirmed that the relationship between flow rate and potential gradient becomes nonlinear at high velocity. Thus, consideration of non-Darcy flow is of importance in hydraulic fractures and near the wellbore.

In 1901, Philippe Forchheimer presented an empirical equation that described the observed nonlinear flow behavior. Forchheimer observed that the deviation from linearity in Darcy's law increased with flow rate. When the fluid velocity increases, for example near the drain inside hydraulic fractures, significant inertial (non-Darcy) effect can occur. This induces an additional pressure drop in the hydraulic fractures to maintain the production rate. Forchheimer attributed the nonlinear increase in pressure gradient to inertial losses in the porous medium, which are proportional to ρv^2. He proposed a second proportionality constant β to describe the increasing contribution to the pressure drop caused by inertial losses. His equation assumes that Darcy's law is still valid, but that an additional pressure drop must be included:

$$-\frac{dp}{dx} = \frac{\mu v}{k} + \beta \rho v^2, \qquad (3.8)$$

where β is the non-Darcy coefficient and ρ the fluid density. The non-Darcy coefficient is usually determined by the analysis of multirate pressure test results. However, such data are not always available so that people have to use correlations obtained from the literature.

Depending on lithology and parameters, there are various theoretical and empirical correlations of the non-Darcy flow in literature (Evans & Civan, 1994; Li & Engler, 2001). Theoretical non-Darcy models can be classified into two groups: parallel-type models and serial-type models. Investigators use capillary models to describe fluid flow through porous media. Scheidegger (1958) and Bear (1988) summarized many researchers' work on capillary models. Parallel-type models assume that a porous medium is made up of a bundle of straight, parallel capillaries of uniform diameter.

Ergun and Orning (1949) argued that the total energy loss for fluid flow through porous media comprises two parts: viscous energy and kinetic energy. Based on the parallel-type model, by combining the Poiseuille equation and the equation similar to the one introduced by Brillouin (1907) for capillary flow, Ergun and Orning developed a theoretical equation to describe the nonlinear laminar flow,

$$-\frac{dp}{dx} = 2\alpha'\frac{(1-\phi)^2}{\phi^3}S_{gv}^2\mu v + \frac{\beta'}{8}\frac{1-\phi}{\phi^3}S_{gv}\rho v^2 \qquad (3.9)$$

where α' and β' are the correction factors, ϕ is the porosity, and S_{gv} is the specific surface defined as the solid surface area divided by the solid volume. Based on the same model, Irmay (1958) theoretically derived Darcy and Forchheimer equations from the dynamic Navier-Stokes equations. When the inertial term is not zero, Irmay derived the following equation,

$$-\frac{dp}{dx} = \frac{\beta''(1-\phi)^2\mu}{\phi^3 D_c^2}v + \frac{\alpha''(1-\phi)\rho}{\phi^3 D_c}v^2 \qquad (3.10)$$

where α'' and β'' are correction factors. Comparing Eqs. (3.9) and (3.10) with the Forchheimer equation, we can find an equation for the non-Darcy coefficient β as follows,

$$\beta = \frac{c}{k^{0.5}\phi^{1.5}}, \qquad (3.11)$$

where c is constant.

A drawback of the parallel models is that all the pores are assumed to go from one face of the porous media to the other. Scheidegger (1958) put forward a serial model in which all the pore space is serially lined up, so that each particle of fluid would have to enter at one pinhole at one side of a porous medium, go through very tortuous channels through all the pores, and then emerge at only one pinhole at the other face of the porous medium. This kind of model is called serial-type model because capillaries of different pore diameter are aligned in series. Assuming a model of length x where there are n capillaries per unit area in each dimensional direction of pore diameter δ and length s, Scheidegger (1958) derived an equation to describe the non-Darcy flow,

$$\frac{dp}{dx} = u\frac{3c\tau^2}{\phi}\mu\left(\int_{\delta_R}^{\infty}\frac{\alpha(\delta)d\delta}{\delta^6}\right)\left(\int\delta^2\alpha(\delta)d\delta\right)^2$$
$$+ u^2\frac{9c'\tau^3}{\phi^2}\rho\left(\int_0^{\delta_R}\frac{\alpha(\delta)d\delta}{\delta^7}\right)\left(\int\delta^2\alpha(\delta)d\delta\right)^3, \qquad (3.12)$$

where τ is the tortuosity, δ_R the critical pore diameter that separates non-Darcy region from Darcy region, $c = 32$, $c' = 1/2$, and $\alpha(\delta)$ the differential pore size distribution function.

Comparing Eq. (3.12) with the Forchheimer equation, the non-Darcy coefficient β can be obtained as follows,

$$\beta = \frac{c''\tau}{k\phi}, \qquad (3.13)$$

where c'' is the constant related to pore size distribution.

Various researchers presented empirical correlations for non-Darcy flow. Based on the theoretical equation developed by Ergun and Orning (1949), Ergun (1952) found an empirical equation by analyzing the data from 640 experiments, which involved various-sized spheres, sand, pulverized coke with several gases such as CO_2, nitrogen, methane, and hydrogen. According to Thauvin and Mohanty's review (1998), the comparison of Ergun's empirical flow equation with the Forchheimer equation leads to

$$\beta = ab^{-0.5}\left(10^{-8}k\right)^{-0.5}\phi^{-1.5}, \qquad (3.14)$$

where $a = 1.75$, $b = 150$, k is expressed in Darcy and β in 1/cm. Macdonald, Elsayed, Mow, and Dullien

(1979) analyzed Eq. (3.9) for particles of different roughness and found that $b = 180$, in the meantime, a ranged from 1.8 to 4.

Janicek and Katz (1955) proposed the equation to predict the non-Darcy coefficient for natural porous media as shown:

$$\beta = 1.82 \times 10^8 k^{-1.25} \phi^{-0.75}, \tag{3.15}$$

where k is expressed in md and β in 1/cm.

Cooke (1973) studied non-Darcy flow for brines, reservoir oils, and gases in propped fractures and used only permeability to predict the non-Darcy coefficient,

$$\beta = bk^{-a}, \tag{3.16}$$

where a and b are constants determined by experiments based on proppant type. Eq. (3.16) is simple and applicable to different types of proppants.

By processing the data measured from unconsolidated media and consolidated media, Geertsma (1974) found that Eq. (3.11) was not applicable to consolidated materials but to unconsolidated media. After he analyzed the data obtained for unconsolidated sandstones, consolidated sandstones, limestones, and dolomites from his and other experiments (Cornell & Katz, 1953; Green & Duwez, 1951), he obtained an empirical correlation,

$$\beta = \frac{0.005}{k^{0.5} \phi^{5.5}}, \tag{3.17}$$

where k is in cm^2 and β in 1/cm.

Pascal and Quillian (1980) proposed a mathematical model to estimate the fracture length and the non-Darcy coefficient. By using the model and the data from variable rate tests from low permeability hydraulically fractured wells, they calculated the non-Darcy coefficients. Based on their results, they proposed an empirical correlation,

$$\beta = \frac{4.8 \times 10^{12}}{k^{1.176}}, \tag{3.18}$$

where k is in md and β in 1/m.

Jones (1987) carried out experiments on 355 sandstone and 29 limestone cores, which were in different core types such as vuggy limestone, crystalline limestone, and fine-grained sandstone. By analyzing the data from experiments, he arrived at a correlation to estimate the non-Darcy coefficient,

$$\beta = \frac{6.15 \times 10^{10}}{k^{1.55}}, \tag{3.19}$$

where k is expressed in md and β in 1/ft.

Liu, Civan, and Evans (1995) plotted Eq. (3.17) developed by Geertsma (1974) against the data obtained, respectively, by Cornell and Katz (1953), Geertsma (1974), Evans, Hudson, and Greenlee (1987), and Whitney (1988). They found that Eq. (3.17) is inaccurate. By considering the effect of tortuosity of the porous medium on the non-Darcy coefficient, they got a better regression fit correlation than Eq. (3.17),

$$\beta = 8.91 \times 10^8 k^{-1} \phi^{-1} \tau, \tag{3.20}$$

where k is expressed in md and β in 1/ft.

Thauvin and Mohanty (1998) developed a pore-level network model to describe high-velocity flow. They input pore size distributions and network coordination numbers into the model, and they obtained outputs such as permeability, non-Darcy coefficient, tortuosity, and porosity. After analyzing all the data they collected, they derived the correlation,

$$\beta = \frac{1.55 \times 10^4 \tau^{3.35}}{k^{0.98} \phi^{0.29}}, \tag{3.21}$$

where k is expressed in Darcy and β in 1/cm.

While suspecting tortuosity may influence the non-Darcy coefficient, by using two different methods to process data from measurements on limestone and sandstone samples, Coles and Hartman (1998) proposed two equations to calculate the non-Darcy coefficient,

$$\beta = \frac{1.07 \times 10^{12} \phi^{0.449}}{k^{1.88}}, \tag{3.22}$$

and

$$\beta = \frac{2.49 \times 10^{11} \phi^{0.537}}{k^{1.79}}, \tag{3.23}$$

where k is expressed in md and β in 1/ft. Comparing Eqs. (3.22) and (3.23) with correlations developed by other researchers, the exponents for porosity in Eqs. (3.22) and (3.23) are positive instead of being negative in other equations.

Cooper, Wang, and Mohanty (1999) conducted non-Darcy flow studies in anisotropic porous media with a microscopic model. They also included tortuosity in predicting the non-Darcy coefficient,

$$\beta = \frac{10^{-3.25} \tau^{1.943}}{k^{1.023}}, \tag{3.24}$$

where k is expressed in cm^2 and β in 1/cm.

Li, Svec, Engler, and Grigg (2001) incorporated non-Darcy effect into a reservoir simulator and simulated the non-Darcy flow experiments, where nitrogen was injected at various flow rates, in several different directions, into a hockey puck-shaped wafer Berea sandstone core sample with a 3-in diameter and a 3/8-in thickness. Comparing differential pressures from simulations with their counterparts from experiments, they found the correlation of β for Berea sandstone,

$$\beta = \frac{11,500}{k\phi}, \quad (3.25)$$

where k is in Darcy and β in 1/cm.

The above equations for predicting the non-Darcy coefficient are valid only for single-phase case. Several researchers conducted non-Darcy flow experiments in multiphase systems, and they obtained some empirical equations to predict the non-Darcy coefficient. In addition to the one-phase correlation of Eq. (3.17), Geertsma (1974) proposed a new β relationship in a two-phase flow. He is the first researcher who addressed non-Darcy flow in the multiphase system. He argued that, in the two-phase system, the permeability in Eq. (3.17) would be replaced by the gas effective permeability at certain water saturation, whereas the porosity would be replaced by the void fraction occupied by the gas. Therefore, in the two-phase system, where the fluids are immobile, the β correlation becomes

$$\beta = \frac{0.005}{k^{0.5}\phi^{5.5}} \frac{1}{(1 - S_{wr})^{5.5}k_r^{0.5}}, \quad (3.26)$$

where S_{wr} is the residual water saturation (or called immobile liquid saturation) and k_r is the gas-relative permeability. The units of k_r and ϕ in Eq. (3.26) are the same as their counterparts in Eq. (3.17). Depending on Eq. (3.26), the presence of the liquid phase increases the non-Darcy coefficient. Wong (1970) also found that β increases by eight times when fluid saturation increases from 40% to 70%. Evans et al. (1987), Grigg and Hwang (1998), and Coles and Hartman (1998) presented consistent results that the non-Darcy coefficient increased with increased liquid saturation.

Based on experimental and analytic investigations, Kutasov (1993) found that Eq. (3.27) could estimate β for both with a mobile liquid saturation and with an immobile liquid saturation.

$$\beta = \frac{1,432.6}{k_g^{0.5}[\phi(1 - S_w)]^{1.5}}, \quad (3.27)$$

where k_g is gas effective permeability expressed in Darcy, β is in 1/cm, and S_w is water saturation.

Frederick and Graves (1994) obtained 407 data points from their experiments, where permeability varied from 0.00197 to 1230 md, and data obtained by Cornell and Katz (1953), Geertsma (1974), and Evans, Marconi, and Tarazona (1986). By using two different regression methods to analyze the data and considering the water saturation effect, Frederick and Graves developed two empirical correlations:

$$\beta = \frac{2.11 \times 10^{10}}{k_g^{1.55}[\phi(1 - S_w)]}, \quad (3.28)$$

and

$$\beta = \frac{1}{[\phi(1 - S_w)]^2}e^{45 - \sqrt{407 + 81 \ln \frac{k_g}{\phi(1 - S_w)}}}, \quad (3.29)$$

where k_g is expressed in md and β is in 1/ft. Although Eqs. (3.28) and (3.29) were achieved from systems with immobile liquid saturation, Frederick and Graves found the two equations can be used in systems with mobile liquid saturation.

Coles and Hartman (1998) conducted non-Darcy experiments with nitrogen and paraffin. They found that β increased with paraffin saturation. When the paraffin saturation was less than 20%, they found that Eq. (3.30) fitted their measurements.

$$\beta = \beta_{dry}e^{6.265S_p}, \quad (3.30)$$

where β_{dry} is the one phase non-Darcy coefficient and S_p is the paraffin saturation.

GAS ADSORPTION

The gas storage mechanisms in shale formations are entirely different from those in conventional gas reservoirs. Natural gas is stored both as an adsorbed phase on the shale matrix and organic materials and as conventional free gas in the porous space. Compared with conventional gas reservoirs, shale gas reservoirs may store a considerable amount of gas as an adsorbed phase (Mengal & Wattenbarger, 2011; Yu & Sepehrnoori, 2014). For instance, in the shale formations of the Appalachian, Michigan, and Illinois basins, several thousand trillion cubic feet of gas is estimated to be contained. Experimental measurements have also indicated that more than 50% of the total gas storage in the Devonian shales may exist as an adsorbed phase (Lu, Li, & Watson, 1995a).

Gas desorption may be a major gas production mechanism and can be an essential factor for ultimate gas recovery. Neglecting the gas desorption effect might lead to underestimating gas potential, especially in shale formations with higher total organic content (TOC). The unprecedented growth of shale reservoirs has brought focus to the investigation of potential contributions of adsorbed gas to the estimated ultimate recovery (EUR) for short-term and long-term periods of production. Some studies have suggested that gas desorption may contribute additional gas production for EUR in shale gas reservoirs (Yu and Sepehrnoori, 2014). Cipolla, Lolon, Erdle, and Rubin (2010) reported that gas desorption might constitute 5%—15% of the total gas production in the 30-year period for both Barnett Shale and Marcellus Shale. However, the impact of gas desorption is primarily observable during the later time of well production, depending on reservoir permeability, flowing bottom hole pressure, and fracture spacing. Thompson, Mangha, and Anderson (2011) observed that gas desorption contributes to 17% increase in the EUR with respect to a 30-year forecasting result in a Marcellus Shale well completed with 12 stages of hydraulic fracturing, located in Northeast Pennsylvania. Mengal and Wattenbarger (2011) presented that gas desorption can result in approximately 30% increase in original gas in place (OGIP) estimates and 17% decrease in recovery factor estimates for Barnett Shale and concluded that it is impossible to obtain

accurate estimations and forecasting if the gas desorption is ignored.

To simulate gas production in shale gas reservoirs, an accurate model of gas adsorption is very important. According to the standard classification system of the International Union of Pure and Applied Chemistry (IUPAC) (Sing, 1982), there are six different types of adsorption, as shown in Fig. 3.3 (Donohue & Aranovich, 1998). The shape of the adsorption isotherm is closely related to the properties of adsorbate and solid adsorbent, and on the pore-space geometry (Silin & Kneafsey, 2012).

The majority of adsorption isotherms may be grouped into the six types as shown in Fig. 3.3, and a detailed description of these isotherms was presented by Sing (1982). In most cases, the isotherm reduces to a linear form at sufficiently low surface coverage, which is often referred to as Henry's law region. Henry's adsorption isotherm presents that the amount of the adsorbate is proportional to the partial pressure. The Type I isotherm is concave to the relative pressure axis and amount of adsorbate approaches limiting constant value. Type I isotherms are given by microporous solids having relatively small external surfaces (e.g., activated carbons, molecular sieve zeolites, and certain porous oxides), the limiting uptake being governed by the accessible micropore volume rather than by the internal surface

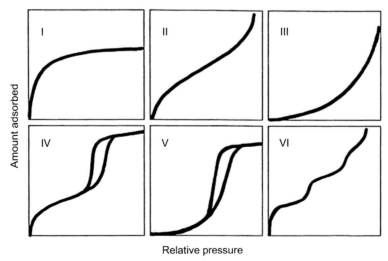

FIG. 3.3 The IUPAC classification of adsorption types. (Credit: Donohue, M. D., & Aranovich, G. L. (1998). Classification of Gibbs adsorption isotherms. *Advances in Colloid and Interface Science, 76*, 137—152. https://doi.org/10.1016/S0001-8686(98)00044-X.)

area. The Type II isotherm is the normal form of isotherm obtained with a nonporous or macroporous adsorbent. The Type II isotherm represents unrestricted monolayer-multilayer adsorption. The beginning of the almost linear middle section of the isotherm is often taken to indicate the stage at which monolayer coverage is complete and multilayer adsorption about to begin. The Type III isotherm is convex to the relative pressure axis over its entire range. Isotherms of this type are not common; the best-known examples are found with water vapor adsorption on pure nonporous carbons. Characteristic features of the Type IV isotherm are its hysteresis loop, which is associated with capillary condensation taking place in mesopores, and the limiting uptake over a range of high pressure. The initial part of the Type IV isotherm is attributed to monolayer-multilayer adsorption because it follows the same path as the corresponding part of a Type II isotherm obtained with the given adsorptive on the same surface area of the adsorbent in a nonporous form. Many mesoporous industrial adsorbents give type IV isotherms. The Type V isotherm is uncommon; it is related to the Type III isotherm in that the adsorbent-adsorbate interaction is weak but is obtained with certain porous adsorbents. The Type VI isotherm represents stepwise multilayer adsorption on a uniform nonporous surface. The step-height represents the monolayer capacity for each adsorbed layer and, in the simplest case, remains nearly constant for two or three adsorbed layers.

In 1918, Irving Langmuir derived a scientific-based adsorption isotherm. The model applies to gases adsorbed on solid surfaces (Langmuir, 1918). The Langmuir adsorption model, which represents Type I of adsorption classification, explains adsorption by assuming an adsorbate behaves as an ideal gas at isothermal conditions. It is also assumed that there is only a single layer of molecules covering the solid surface, as shown in Fig. 3.4A. It is a semiempirical isotherm with a kinetic basis and was derived based on statistic thermodynamics. The general form of Langmuir isotherm is presented as follows:

$$V = \frac{V_L p}{p_L + p},\qquad(3.31)$$

where V is the adsorbed gas volume at pressure p, V_L the Langmuir volume, which is maximum volume of adsorption at the infinite pressure, and p_L the Langmuir pressure, which is the pressure corresponding to one-half Langmuir volume presented in Fig. 3.5. Inherent within this model, the several assumptions are considered specifically for the simplest case. All of the adsorption sites are equivalent, and each site can only accommodate one molecule. The surface is energetically homogeneous, and adsorbed molecules do not interact. Adsorbed gas is immobile. At the maximum adsorption, only a monolayer is formed. Adsorption only occurs on localized sites on the surface, not with other adsorbates. In reality, these assumptions are rarely

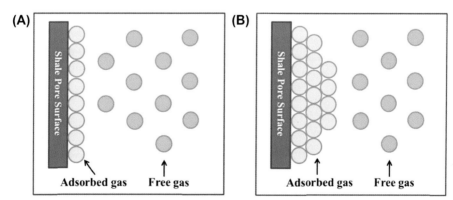

FIG. 3.4 The schematic plots of **(A)** monolayer and **(B)** multilayer gas adsorption. (Credit: Yu, W., Sepehrnoori, K., & Wiktor Patzek, T. (2014). Evaluation of gas adsorption in Marcellus shale. *Paper presented at the SPE annual technical conference and exhibition, Amsterdam, The Netherlands.* https://doi.org/10.2118/170801-MS.)

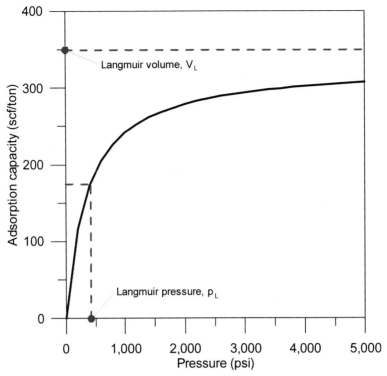

FIG. 3.5 Typical Langmuir isotherm curve.

true. Surfaces are always heterogeneous, and adsorbed molecules are not necessarily immobile. In addition, frequently more molecules will adsorb to the monolayer. Nevertheless, the Langmuir isotherm is the first choice for most models of adsorption and has many applications in most shale reservoir cases.

In 1938, Stephen Brunauer, Paul Hugh Emmett, and Edward Teller published the article about the BET theory to explain the physical multilayer adsorption of gas molecules on the solid surface as shown in Fig. 3.4B. The concept of the BET isotherm model is an extension of the Langmuir model, which is a model for monolayer adsorption to multilayer adsorption. In BET theory, gas molecules physically adsorb on a solid in layers infinitely, and the concept of Langmuir theory can be applied to each layer. The expression of BET isotherm is shown as follows:

$$V = \frac{V_m C p}{(p_o - p)\left[1 + \frac{(C-1)p}{p_o}\right]}, \quad (3.32)$$

where V_m is the maximum adsorption gas volume when the entire adsorbent surface is being covered with a complete monomolecular layer, C a constant related to the net heat of adsorption, and p_o the saturation pressure of the gas. C is defined as below:

$$C = \exp\left(\frac{E_1 - E_L}{RT}\right), \quad (3.33)$$

where E_1 is the heat of adsorption for the first layer, E_L the heat of adsorption for the second and higher layers and is equal to the heat of liquefaction, R the gas constant, and T the temperature. The assumptions in the BET theory include homogeneous surface, no lateral interaction between molecules, and the uppermost layer is in equilibrium with the gas phase. A more convenient form of the BET adsorption isotherm equation is as follows:

$$\frac{p}{V(p_o - p)} = \frac{1}{V_m C} + \frac{C-1}{V_m C}\frac{p}{p_o} \quad (3.34)$$

A plot of $\frac{p}{V(p_o-p)}$ against $\frac{p}{p_o}$ should give a straight line with an intercept of $\frac{1}{V_m C}$ and slope of $\frac{C-1}{V_m C}$.

The standard BET isotherm assumes that the number of adsorption layers is infinite. In the case of n adsorption layers in some finite number, then a general form of BET isotherm is given below:

$$V(p) = \frac{V_m C \frac{p}{p_o}}{1-\frac{p}{p_o}}\left[\frac{1-(n+1)\left(\frac{p}{p_o}\right)^n+n\left(\frac{p}{p_o}\right)^{n+1}}{1+(C-1)\frac{p}{p_o}-C\left(\frac{p}{p_o}\right)^{n+1}}\right], \quad (3.35)$$

where n is the maximum number of adsorption layers. When $n=1$, Eq. (3.35) will be reduced to the Langmuir isotherm, Eq. (3.31). When $n=\infty$, Eq. (3.35) will be reduced to Eq. (3.32). According to Yu, Sepehrnoori, and Patzek (2016), at high reservoir pressures, natural gas adsorbed on the organic carbon surfaces forms multimolecular layers. They observed that the gas desorption in some areas of the Marcellus Shale follows the BET isotherm on the basis of laboratory experiments. In other words, the Langmuir isotherm may not be a good approximation for the amount of gas adsorbed on shale reservoir, and the BET isotherm should be a better choice.

In an effort to improve understanding of adsorption in shale reservoirs, numerous adsorption experiments were carried out on shale rocks (Heller & Zoback, 2014; Lu, Li, & Watson, 1995b; Ross & Bustin, 2007; Nuttall et al., 2005; Vermylen, 2011). They measured the surface uptake of an adsorbate depending on various pressures at a constant temperature to quantify the adsorptive potential of a material, defining an adsorption isotherm. The magnitude and shape of the adsorption isotherm give a better understanding of pore structure and surface properties of the material. Procedures for measuring adsorption are well established and can generally fall into the category of either mass-based or volumetric-based methods (Heller & Zoback, 2014). Mass-based methods, which are commonly used in material science, directly measure the change of sample mass associated with adsorption. The advantage of mass-based methods is a very high degree of accuracy, with the trade-off being the need to use tiny sample sizes. In the oil and gas industry, volumetric method, which allows us to use much larger sample volumes, has been generally used. Volumetric method is similar to porosity measurement based on Boyle's law. Void volume measurements were performed with both adsorbing and nonadsorbing gas to calculate the amount of adsorption.

Heller and Zoback (2014) measured CH_4 and CO_2 adsorption isotherms on samples from Barnett, Eagle Ford, Marcellus, and Montney reservoirs (Fig. 3.6). In Fig. 3.6, all data were fit to Langmuir isotherms. Especially, CO_2 presents a greater capacity for adsorption compared with CH_4. In this experiment, CO_2 adsorbed two times greater in Barnett and Marcellus samples, and three times greater in the Eagle Ford samples. Several research reports have shown that the affinity of CO_2 adsorption to the shale reservoir is greater than that of CH_4 under subsurface conditions, depending on the thermal maturity of the organic materials (Busch et al., 2008; Kim, Cho, & Lee, 2017; Shi & Durucan, 2008). Consequently, interest in CO_2 injection for CO_2 storage and enhanced gas recovery (EGR) in shale reservoirs has grown recently. CO_2 injection in shale reservoirs will be described in detail in Chapter 5. Although most literature as well as Heller and Zoback (2014) argue that the monolayer Langmuir isotherm describes adsorption behavior in shale reservoirs, Yu et al. (2016) presented that measurements of methane adsorption deviate from the Langmuir isotherm. In four Marcellus Shale core samples, adsorption isotherms obey not the Langmuir theory but the BET theory as shown in Fig. 3.7. In these experiments, data correspond with Langmuir adsorption at low pressures. However, at high reservoir pressures, all data deviate from Langmuir isotherms and fit better with BET isotherms. Most previous researches performed adsorption experiments at low pressure so that behavior of adsorption at high pressure should be analyzed in detail. In addition, according to Vermylen (2011), CO_2 adsorption also obeys BET isotherm.

Recently, the effects of adsorption on shale permeability have been studied by several researchers. Measuring permeability of shale samples, Sinha et al. (2013) and Cao et al. (2016) found that gas adsorption affects the permeability of shale. Although the effect of adsorption on permeability has not been thoroughly understood for shale, there are a few studies on adsorption effects for coal. Palmer and Mansoori (1998) presented a theoretical model for calculating pore volume compressibility and permeability in coals as a function of pressure and matrix shrinkage. They introduced porosity and permeability multipliers as follows:

$$\frac{\phi}{\phi_i} = 1+\frac{c_f}{\phi_i}(p-p_i)+\frac{\varepsilon_l}{\phi_i}\left(1-\frac{K}{M}\right)\left(\frac{p_i}{p_i+p_L}-\frac{p}{p+p_L}\right), \quad (3.36)$$

$$\frac{k}{k_i} = \left(\frac{\phi}{\phi_i}\right)^n, \quad (3.37)$$

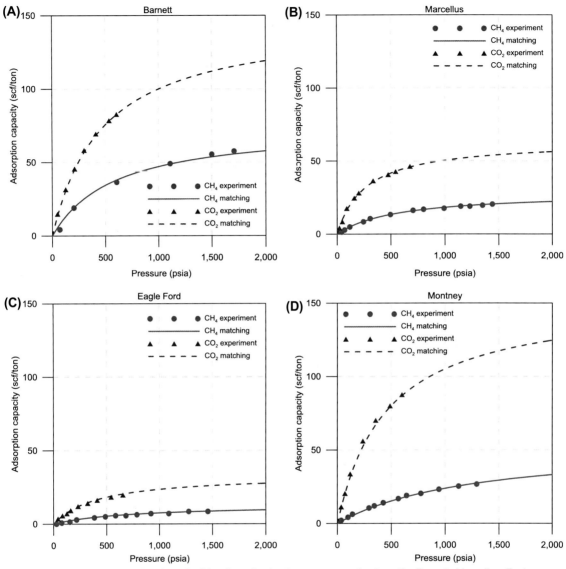

FIG. 3.6 Methane and carbon dioxide adsorption isotherms on samples from the Barnett, Marcellus, Eagle Ford and Montney shale reservoirs. (Credit: Reproduced from Heller, R., & Zoback, M. (2014). Adsorption of methane and carbon dioxide on gas shale and pure mineral samples. *Journal of Unconventional Oil and Gas Resources, 8*, 14–24. https://doi.org/10.1016/j.juogr.2014.06.001.)

where

$$\frac{K}{M} = \frac{1}{3}\left(\frac{1+\nu}{1-\nu}\right), \qquad (3.38)$$

c_f is the pore volume compressibility, ε_l the strain at infinite pressure, K the bulk modulus, and M the axial modulus. The third term of Eq. (3.36) indicates strain

change depending on adsorption effects. To couple the effects of adsorption and stress change on permeability, the term was combined with Eq. (3.31) as shown below (Kim, 2018):

$$k = k_i\left\{\left(\frac{\sigma'}{\sigma'_i}\right)^{-b} + \left[\frac{\varepsilon_l}{\phi_i}\left(1-\frac{K}{M}\right)\left(\frac{p_i}{p_i+p_L}-\frac{p}{p+p_L}\right)\right]^n\right\}.$$

$$(3.39)$$

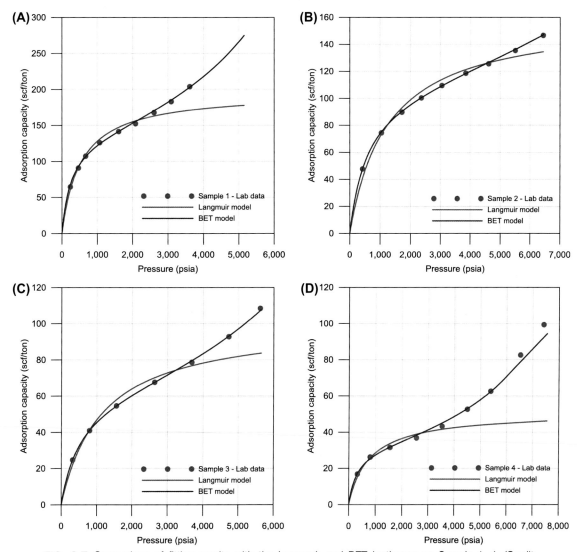

FIG. 3.7 Comparison of fitting results with the Langmuir and BET isotherms on Sample 1–4. (Credit: Reproduced from Yu, W., Sepehrnoori K., & Patzek, T. W. (2016). Modeling gas adsorption in Marcellus shale with Langmuir and BET isotherms. *SPE Journal, 21*(2), 589–600. https://doi.org/10.2118/170801-Pa.)

Kim (2018) matched Eq. (3.39) with experimental data of Guo, Hu, Zhang, Yu, and Wang (2017). Guo et al. (2017) performed adsorption and permeability measurement of shale cores in reservoir condition. They eliminated the influence of stress sensitivity on shale permeability measurement to focus on effects of

adsorption to shale permeability. Fig. 3.8 shows measured CH_4 adsorption and permeability data and fitting results with Langmuir isotherm and Eq. (3.39) in shale cores. Table 3.1 shows the input value for permeability fitting.

Based on Eq. (3.39), which considered strain change depending on Langmuir isotherm, Kim (2018)

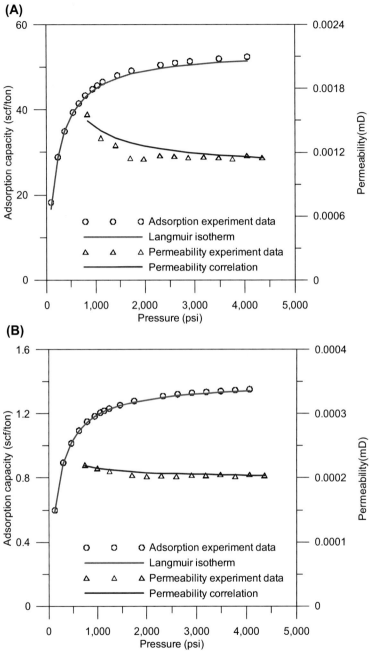

FIG. 3.8 Measured CH_4 adsorption and permeability data and fitting results with Langmuir isotherm and permeability correlation in shale cores.

TABLE 3.1

Fitting Values for Langmuir Adsorption and Permeability Correlation in Shale Cores

Fitting Parameters	Core A	Core B
Langmuir pressure	201	174
Langmuir volume	54	49
Strain at infinite pressure	0.04	0.03
Ratio of bulk to axial modulus	0.5	0.5

extended strain-dependent permeability correlation combining with BET isotherm (Eq. 3.35) as follows:

$$
k = k_i \left\{ \left(\frac{\sigma'}{\sigma'_i} \right)^{-b} + \frac{\varepsilon_l}{\phi_i} \left(1 - \frac{K}{M} \right) \right.
$$

$$
\left(\frac{C \frac{p_i}{p_o} \left[1 - (n+1)\left(\frac{p_i}{p_o}\right)^n + n\left(\frac{p_i}{p_o}\right)^{n+1} \right]}{1 - \frac{p_i}{p_o}} \cdot \frac{1}{1 + (C-1)\frac{p_i}{p_o} - C\left(\frac{p_i}{p_o}\right)^{n+1}} \right) \quad (3.40)
$$

$$
\left. - \frac{C \frac{p}{p_o} \left[1 - (n+1)\left(\frac{p}{p_o}\right)^n + n\left(\frac{p}{p_o}\right)^{n+1} \right]}{1 - \frac{p}{p_o}} \cdot \frac{1}{1 + (C-1)\frac{p}{p_o} - C\left(\frac{p}{p_o}\right)^{n+1}} \right\}.
$$

Permeability change resulting from stress change and adsorption effects can be calculated by Eqs. (3.39) and (3.40).

NANO SCALE FLOW

Shale reservoirs present various unconventional features and considerations, in particularly on nanoscale of flow. Nanoscale flow of hydrocarbons through porous media involves various distinct transport mechanisms (Freeman, Moridis, & Blasingame, 2011). The flow of gases through microporous zeolites (Hassan & Douglas Way, 1996) and other nanoporous materials (Tzoulaki et al. 2009) were studied. Many of the insights obtained from works on micro- and nanoscale transport are applicable to transport through shale. Unlike conventional reservoirs, the migration of shale gas is strongly influenced by pore size and properties of porous media. Therefore, the conventional continuum flow equation, Darcy equation, is no longer applicable for describing complex gas transport mechanisms in

nanopores of shale reservoirs. Various flow mechanisms in shale reservoirs have been studied recently (Akkutlu, Efendiev, & Savatorova, 2015; Civan, 2010; Fathi, Tinni, & Akkutlu, 2012; Javadpour, 2009; Javadpour, Fisher, & Unsworth, 2007; Moghanloo et al., 2015; Sheng et al., 2015; Sheng, Javadpour, & Su, 2018; Wu, Li, Wang, Chen, & Yu, 2015; Zheng, Yuan, Rouzbeh, & Moghanloo, 2017). Among potential transport mechanisms in shale reservoirs, which differ from those in conventional reservoirs, slippage and Knudsen diffusion, which are essential in the shale gas flow, will be introduced in this section.

The pores in shale reservoirs are in the range of 1−100 nm so that the size of gas molecules contained in the pores (∼0.5 nm) is comparable with pore size. Under certain pressure and temperature conditions, the distance between hydrocarbon molecules (mean free path) exceeds the size of the pores. In such circumstances, the gas molecules might move singly through the pores, and the flow behavior in nanopores deviates from the concept of continuum and bulk flow (Fig. 3.9). To describe these flow behaviors in nanoscale pores of shale reservoirs, several scholars have introduced the concept of apparent permeability.

Javadpour (2009) first introduced an apparent permeability model that couples convective flow and Knudsen diffusion in nanopores by comparing formulation for gas flow with the Darcy equation. On the basis of the Javadpour model, many scholars have developed numerous shale models that can depict complex gas transport mechanisms by apparent permeability. In these days, there are representative types of apparent permeability models for shale gas reservoirs (Sheng et al., 2018): Javadpour models are based on the pore radius of porous media (Akkutlu et al., 2015; Akkutlu & Fathi, 2012; Azom & Javadpour, 2012; Darabi, Ettehad, Javadpour, & Sepehrnoori, 2012; Javadpour, 2009; Javadpour et al., 2007; Sheng et al., 2018, 2015; Singh & Javadpour, 2016; Wasaki & Akkutlu, 2015; Zhang et al., 2015), and Civan models are based on Knudsen number (Civan, Devegowda, & Sigal, 2013; Civan, Rai, & Sondergeld, 2011; Islam & Patzek, 2014; Song et al., 2016; Wang et al., 2017; Wu et al., 2015; Yuan, Wood, & Yu, 2015). Javadpour models characterize intrinsic permeability, Knudsen diffusion coefficient, and slip factor using pore size and propose coupled flow equations considering viscous flow, slippage effect, Knudsen diffusion, and surface diffusion. Civan models apply the Beskok and Karniadakis model (1999) to describe gas migration in porous media and use the Knudsen number in models to couple viscous flow and Knudsen diffusion. In the following section,

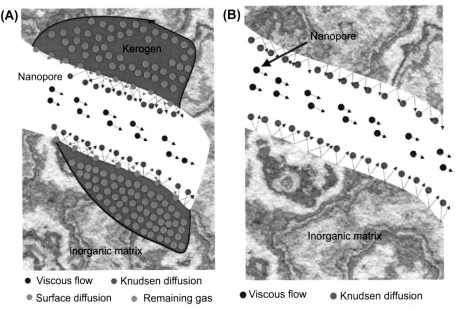

FIG. 3.9 Shale gas flow in nanoscale pores of organic and inorganic materials. (Credit: Sheng, G., Javadpour, F., & Su, Y. (2018). Effect of microscale compressibility on apparent porosity and permeability in shale gas reservoirs. *International Journal of Heat and Mass Transfer, 120*, 56–65. https://doi.org/10.1016/j. ijheatmasstransfer.2017.12.014.)

detailed characteristics of flow mechanisms in shale reservoir will be described.

Knudsen number, K_n, is the ratio of the mean free path, λ, to pore diameter, d, and can be used to identify different flow regimes in the porous media as given below:

$$K_n = \frac{\lambda}{d}, \tag{3.41}$$

where

$$\lambda = \frac{k_B T}{\sqrt{2}\pi\delta^2 p}, \tag{3.42}$$

in which k_B is the Boltzmann constant ($1.3806488 \times 10^{-23}$ J/K), and δ is the collision diameter of the gas molecule. According to several studies (Rathakrishnan, 2004; Rezaee, 2015; Roy, Raju, Chuang, Cruden, & Meyyappan, 2003), gas transport can be divided into different flow patterns by Knudsen number. In each flow regime, gas transport follows different flow equations. For $K_n < 10^{-3}$, the gas transport in pores is continuous flow without slippage effect. In this flow regime, gas can be considered as a continuous medium and gas flow conforms to Darcy equation. For $10^{-3} < K_n < 10^{-1}$, the gas transport in pores is continuous flow with slippage effect. In this flow regime, gas still can be considered as a continuous medium so that gas flow meets Darcy's law. In addition, gas flow along the pore wall is not zero, and slippage effect exists so that gas flow reaches the Klinkenberg's equation. Therefore, the gas flow in the regime is influenced by both Darcy and slippage effect. For $10^{-1} < K_n < 10$, the gas transport in pores is transitional flow. In this regime, λ and d are in the same order of magnitude, and the influence of collisions between gas molecules on gas flow is as important as collisions between gas molecules and pore wall. Because the assumption of the continuum flow is no longer valid, gas transport in this regime is the combination of the Knudsen diffusion and slip flow. For $K_n > 10$, the gas transport in pores is free-molecule flow. In this flow regime, collisions between gas molecules are no longer critical, collisions between gas molecules and the pore wall become the main affecting factor, and gas transport in this regime only meets the Knudsen diffusion. Fig. 3.10 shows the summary of various flow regimes and control equations depending on Knudsen number.

When K_n is in the range of 0.001 and 0.1, the collision between the gas molecule and the internal surface wall cannot be ignored, and the slippage effect is not

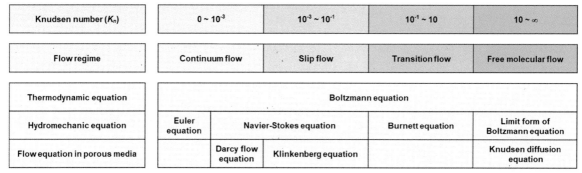

Knudsen number (K_n)		$0 \sim 10^{-3}$	$10^{-3} \sim 10^{-1}$	$10^{-1} \sim 10$	$10 \sim \infty$
Flow regime		Continuum flow	Slip flow	Transition flow	Free molecular flow
Thermodynamic equation		Boltzmann equation			
Hydromechanic equation	Euler equation	Navier-Stokes equation		Burnett equation	Limit form of Boltzmann equation
Flow equation in porous media		Darcy flow equation	Klinkenberg equation		Knudsen diffusion equation

FIG. 3.10 Various flow regimes and control equations depending on Knudsen number.

negligible. Therefore, apparent gas permeability is higher than that measured by liquid. Klinkenberg (1941) showed experimentally that a linear relationship exists between Darcy permeability and the reciprocal of mean pressure in the system as shown:

$$k_g = k_D \left(1 + \frac{b}{p_{\text{avg}}} \right), \qquad (3.43)$$

where k_g is the gas permeability at mean pressure, p_{avg}, k_D the Darcy permeability or liquid permeability, and b the Klinkenberg parameters. The experimental parameters b and k_D are the slope and intercept of the fitted line through the k_g versus $\frac{1}{p_{\text{avg}}}$ data (Fig. 3.11).

Javadpour (2009) presented a model including the Knudsen diffusion and the slip flow, which are significant mechanisms contributing to the gas flow in a single, straight, cylindrical nanotube. Javadpour (2009) indicated that total mass flux of a gas through a nanopore is the result of a combination of Knudsen diffusion and pressure forces, as follows:

$$J = J_a + J_D, \qquad (3.44)$$

where J is the total mass flux, J_a the advective mass flux due to pressure forces, and J_D the Knudsen diffusive mass flux. Advective mass flux J_a for an ideal gas in laminar flow in a circular tube can be derived from Hagen-Poiseuille's equation (Bird, Stewart, & Lightfoot, 2007). Javadpour (2009) presented the advective mass flux equation considering the length of the pore as shown below:

$$J_a = -\frac{r^2}{8\mu} \frac{p_{\text{avg}}}{L} \Delta p. \qquad (3.45)$$

For nanoscale pores, the no-slip boundary condition is invalid (Brown, Dinardo, Cheng, & Sherwood, 1946; Hadjiconstantinou, 2006; Hornyak, Tibbals, Dutta, & Moore, 2008; Javadpour et al., 2007; Karniadakis, Beskok, & Aluru, 2005). Slip velocity on the surface of

a nanopore eases the gas flow. Brown et al. (1946) introduced a theoretical dimensionless coefficient F to correct for slip velocity in tubes as follows:

$$F = 1 + \left(\frac{8\pi RT}{M} \right)^{0.5} \frac{\mu}{p_{\text{avg}} r} \left(\frac{2}{\alpha} - 1 \right), \qquad (3.46)$$

where α is the tangential momentum accommodation coefficient or the part of gas molecules reflected diffusely from the tube wall relative to specular reflection (Maxwell, 1995). The value of α varies theoretically in a range from 0 to 1, depending on wall surface smoothness, gas type, temperature, and pressure. Experimental measurements are needed to determine α for specific mudrock systems. Eq. (3.46) shows that smaller pores result in higher values for the multiplier F. Lower pressures also result in higher F.

Roy et al. (2003) showed that Knudsen diffusion in nanopores could be written in the form of the pressure gradient. Experimental steady-state data demonstrate that the relationship between flow rate and pressure drop is linear. Gas mass flux by diffusion with negligible viscous effects in a nanopore is described as follows (Javadpour, 2009; Roy et al., 2003):

$$J_D = \frac{M D_K}{10^3 RT} \nabla p, \qquad (3.47)$$

where M is molar mass, D_K the Knudsen diffusion constant, R the gas constant ($= 8.314$ J/mol/K), and T the absolute temperature in Kelvin. The Knudsen diffusion constant, D_K, is defined as (Javadpour et al., 2007):

$$D_K = \frac{2r}{3} \left(\frac{8RT}{\pi M} \right)^{0.5}, \qquad (3.48)$$

From Eqs. (3.44–3.48), the total mass flux through a nanopore considering a combination of slip flow and Knudsen diffusion is calculated shown below:

$$J = - \left[\frac{2rM}{3 \times 10^3 RT} \left(\frac{8RT}{\pi M} \right)^{0.5} + F \frac{r^2 p_{\text{avg}}}{8\mu} \right] \frac{(p_2 - p_1)}{L}, \qquad (3.49)$$

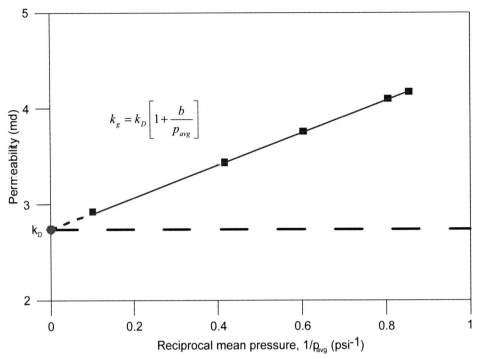

FIG. 3.11 The Klinkenberg effect in gas permeability measurements.

According to Javadpour (2009), Eq. (3.49) shows a reasonable match with experimental data for homogeneous porous media consisting of relatively cylindrical and straight nanopores from Roy et al. (2003).

Volumetric flux based on the Darcy equation for compressible gases used for conventional systems and volumetric flux for nanopores from Eq. (3.49) is calculated as follows:

$$\frac{q}{A} = -\frac{k_D}{\mu} \frac{(p_2 - p_1)}{L}, \tag{3.50}$$

$$\frac{q}{A} = -\left[\frac{2rM}{3 \times 10^3 RT\rho_{avg}}\left(\frac{8RT}{\pi M}\right)^{0.5} + F\frac{r^2}{8\mu}\right]\frac{(p_2 - p_1)}{L}. \tag{3.51}$$

By increasing the pore size or increasing pressure, Eq. (3.51) can be simplified to Darcy equation. Comparing Eqs. (3.50) and (3.51), apparent permeability, k_{app} for the gas flow in mudrock systems is obtained as:

$$k_{app} = \frac{2r\mu}{3 \times 10^3 p_{avg}}\left(\frac{8RT}{\pi M}\right)^{0.5} + \frac{r^2}{8}\left[1 + \left(\frac{8\pi RT}{M}\right)^{0.5}\frac{\mu}{p_{avg}r}\left(\frac{2}{\alpha} - 1\right)\right]. \tag{3.52}$$

According to Eq. (3.52), permeability in the nanopore system is depends not only on rock properties but also on flowing gas properties at specified pressure and temperature. In this equation, Knudsen diffusion, which is negligible for conventional systems, plays an essential role in fine-grained mudrocks.

Based on the results of Javadpour (2009), various modifications of apparent permeability were presented. Because the model of Javadpour (2009) is strictly valid only for the ideal gas condition, Azom and Javadpour (2012) showed modified apparent permeability for a real gas flowing in a porous medium given below:

$$k_{app} = \frac{2r\mu c_g}{3 \times 10^3}\left(\frac{8ZRT}{\pi M}\right)^{0.5} + \frac{r^2}{8}\left[1 + \left(\frac{8\pi RT}{M}\right)^{0.5}\frac{\mu}{p_{avg}r}\left(\frac{2}{\alpha} - 1\right)\right], \tag{3.53}$$

where c_g is gas compressibility and Z is compressibility factor. In Eqs. (3.52) and (3.28), the real gas becomes ideal gas because of the gas compressibility $c_g = \frac{1}{p_{avg}}$ and the compressibility factor $Z = 1$ for an ideal gas.

Darabi et al. (2012) applied several modifications to upscale Javadpour (2009) model from a single straight cylindrical nanotube to ultratight natural porous media, characterized by a network of interconnected tortuous micropores and nanopores. In the Knudsen diffusion term, the porosity/tortuosity factor, $\frac{\phi}{\tau}$, is introduced to model Knudsen flow through porous media (Javadpour et al., 2007). In addition, the fractal dimension of the pore surface, D_f, is included to consider the effect of pore-surface roughness on the Knudsen diffusion coefficient (Coppens, 1999; Coppens & Dammers, 2006). Surface roughness is one example of local heterogeneity. Increasing surface roughness leads to an increase in residence time of molecules in porous media and a decrease in Knudsen diffusivity. D_f is a quantitative measure of surface roughness that varies between 2 and 3, representing a smooth surface and a space-filling surface, respectively (Coppens & Dammers, 2006). The final form of apparent permeability model of Darabi et al. (2012) is

$$k_{app} = \frac{\mu\phi}{\tau p_{avg}}(\delta_r)^{D_f - 2} D_k + \frac{r^2}{8}\left[1 + \left(\frac{8\pi RT}{M}\right)^{0.5}\frac{\mu}{p_{avg}r}\left(\frac{2}{\alpha} - 1\right)\right],$$

(3.54)

where δ' is the ratio of normalized molecular size, d_m, to local average pore diameter, d_p.

In the models above, the value of the tangential momentum accommodation coefficient is used. Tangential momentum accommodation coefficient is one empirical parameter that is necessary to account for slip flow of the gas molecule at the pore wall. The most significant limitation in these models is the estimation of the tangential momentum accommodation coefficient, which requires expensive experiments or molecular dynamic simulations (Agrawal & Prabhu, 2008).

Although the determination of tangential momentum accommodation coefficient in simplified conventional systems has been studied extensively using experimental and numerical approaches, tangential momentum accommodation coefficient data are unavailable for shale reservoirs due to the diversity of organic materials and mineral types, as well as different gas components. Hence, Singh, Javadpour, Ettehadtavakkol, and Darabi (2014) presented a model that requires no empirical coefficients, named as nonempirical apparent permeability (NAP), to reliably predict apparent permeability in shale reservoirs. Singh et al. (2014) derived apparent permeability on the basis of fundamental flow equations for shale gas systems. From the total mass flow, which is a superposition of advection and molecular spatial diffusion (Veltzke & Thoming, 2012), Darcy's law can be converted to expressions for apparent permeability of slits or tubes:

$$(k_{app})_{slit} = \frac{\phi\mu h}{3\tau}\left(\frac{h_{slit}Z}{4\mu} \frac{8}{\pi p_{avg}M}\sqrt{\frac{2MRT}{\pi}}\right)$$

(3.55)

$$(k_{app})_{tube} = \frac{2\phi\mu d}{\pi\tau}\left(\frac{\pi d_{tube}Z}{64\mu} \frac{1}{3 p_{avg}M}\sqrt{2\pi MRT}\right)$$

(3.56)

where h_{slit} is the height of the rectangular slit and d_{tube} the diameter of the tube. The two pore geometries considered in the NAP model are cylindrical tube and rectangular channel (slit). When porous media are composed of other shapes, the permeability of the media will be somewhere between what it would be if it were composed of tubes and what it would be if it were composed of slits. Therefore, the two shapes considered in the NAP model may reliably capture the average effect of different pore shapes in porous media because capturing the exact shape of each pore might be impractical and daunting. The permeability of each shape type contributes to the effective permeability of the reservoir. The effective permeability is the statistic sum of the individual permeability from each shape type (Fenton, 1960) as given below:

$$\ln(k_{app})_{eff} = \frac{x}{100}\ln(k_{app})_{slit} + \frac{100 - x}{100}\ln(k_{app})_{tube}$$

(3.57)

$$(k_{app})_{eff} = \left[\left(k_{app}^{\frac{x}{100}}\right)_{slit}\left(k_{app}^{\frac{100-x}{100}}\right)_{tube}\right]$$

(3.58)

where k_{eff} is effective permeability after including the effect of sorption. Although there are some observed values of pure gases and solid materials in literature, finding them for the shale system is not straightforward. Hence, a method that does not need the empirical value is attractive.

Singh and Javadpour (2016) proposed another apparent permeability model of shale micro/nanopores considering the Langmuir slip condition (Myong, 2001, 2004), which accounts for gas transport due to viscous flow, slip flow, Knudsen diffusion, and sorption. They called this permeability model as Langmuir slip permeability (LSP), and it is valid under slip flow and transition flow regimes. Langmuir slip condition considers the gas-surface molecular interaction, and it originates from the theory of surface chemistry (Adamson & Gast, 1997). Singh and Javadpour (2016) suggested several novelties of LSP model compared with previous

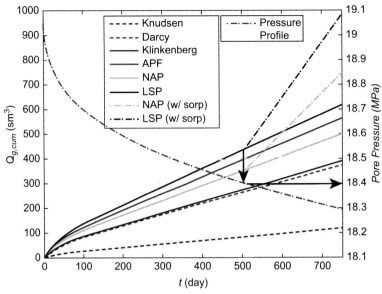

FIG. 3.12 Prediction of cumulative gas production for the LSP, NAP, APF, Klinkenberg, Darcy flow, and Knudsen diffusion models. (Credit: Singh, H., & Javadpour, F. (2016). Langmuir slip-Langmuir sorption permeability model of shale. *Fuel, 164*, 28–37. https://doi.org/10.1016/j.fuel.2015.09.073.)

literature. LSP model compensates shortcoming of Maxwell slip condition. In Maxwell slip condition, the value of tangential momentum accommodation coefficient, which is difficult to obtain in shale rock, should be estimated, and accurate estimation of velocity gradient is not readily available. In addition, Maxwell slip condition cannot explain the difference between the type of gas and surface molecules. Langmuir adsorption data in LSP model, which are easily measured compared with tangential momentum accommodation coefficient, are used to determine slip coefficient for gas flow. LSP model includes the higher-order slip effects on gas flow. Singh and Javadpour (2016) presented Fig. 3.12, which compares results of the LSP, NAP, apparent permeability function (APF), Klinkenberg, Darcy flow, and Knudsen diffusion models.

Langmuir slip model is based on the theory of adsorption, which determines the amount of adsorbed gas as a function of pressure and temperature. Hence, the Langmuir slip model is derived using Langmuir adsorption isotherm (1918). The analytic solution for the flow from the Langmuir slip model is obtained using the approach adopted by Arkilic (1997), which is through the momentum equation. Average flow velocity inside a slit is acquired by the dimensionless velocity from the Langmuir slip model (Myong, 2004). The

average flow velocity is then used in Darcy's law to obtain an apparent permeability as follows:

$$k_{\text{app}} = -\frac{\phi}{\tau} \frac{h^2}{4\delta} \frac{dP}{dX}\left(\frac{2}{3} + \frac{1}{\sqrt{\overline{\beta}P}}\right), \tag{3.59}$$

where

$$\delta = -\left(\frac{dP}{dX}\right)_{X=1}\left(2 + \frac{3}{\sqrt{\overline{\beta}}}\right), \tag{3.60}$$

$$P = \frac{p - p_{\text{min}}}{p_{\text{max}} - p_{\text{min}}}, \tag{3.61}$$

where X is a dimensionless length and $\overline{\beta} = \beta p$ is a dimensionless Langmuir adsorption constant. $\left(\frac{dP}{dX}\right)_{X=1}$ represents the x derivative of the pressure profile at the exit condition or pore outlet.

Civan (2010) developed the apparent permeability model using slip flow assumption, represented by simplified second-order slip model. They improved a unified Hagen-Poiseuille-type equation from Beskok and Karniadakis (1999) covering the fundamental flow regimes in tight porous media. The empirical correlation of the available data of the dimensionless rarefaction coefficient from Beskok and Karniadakis (1999) is a mathematically complicated trigonometric function. Civan (2010) presented a simple inverse-power-law function, which shows much definite correlation with the same data. In addition, they derived apparent permeability model considering several issues such as number and tortuosity of the preferential flow paths in tight porous media. A unified Hagen-Poiseuille-type equation for volumetric gas flow through a single pipe is given below (Beskok & Karniadakis, 1999):

$$q_h = f(K_n) \frac{\pi r^4}{8\mu} \frac{\Delta p}{L_h}, \tag{3.62}$$

where

$$f(K_n) = (1 + \alpha K_n)\left(1 + \frac{4K_n}{1 - bK_n}\right), \tag{3.63}$$

where L_h is the hydraulic length or the length of the tortuous flow paths, α the dimensionless rarefaction coefficient, which varies in the range of $0 < \alpha < \alpha_o$ over $0 \le K_n < \infty$, and b the slip coefficient. The empirical correlation of dimensionless rarefaction coefficient α provided by Beskok and Karniadakis (1999) is given by:

$$\alpha = \alpha_o \frac{2}{\pi} \tan^{-1}\left(\alpha_1 K_n^{\alpha_2}\right). \tag{3.64}$$

Because of a mathematically complicated form of above equation, Beskok and Karniadakis (1999) provided a simple inverse power law expression as given below:

$$\frac{\alpha_o}{\alpha} - 1 = \frac{A}{K_n^B}, \tag{3.65}$$

where A and B are empirical fitting constants. It is demonstrated that simple correlation offers a much more accurate and practical. Civan (2010) derived the volumetric gas flow through a bundle of tortuous flow paths from the unified Hagen-Poiseuille-type equation for flow through a single pipe shown as:

$$q = nq_h = nf(K_n) \frac{\pi r^4}{8\mu} \frac{\Delta p}{L_h}, \tag{3.66}$$

where n denotes the number of preferential hydraulic flow paths formed in porous media. n can be approximated by rounding the value calculated by the following equation to the nearest integer (Civan, 2007):

$$n = \frac{\phi A_b}{\pi r^2}, \tag{3.67}$$

where A_b is the bulk surface area of porous media perpendicular to flow direction. It can be described macroscopically by a Darcy-type gradient-law of flow where the flow is assumed proportional to the pressure gradient given by:

$$q = \frac{k_{app} A_b}{\mu} \frac{\Delta p}{l}, \tag{3.68}$$

Combining above equations, apparent permeability was derived as the following expression:

$$k = \frac{\phi r^2}{8\tau_h} f(K_n), \tag{3.69}$$

where

$$\tau_h = \frac{L_h}{L}. \tag{3.70}$$

In addition to the aforementioned models, there were a lot of studies for nanoscale flow in shale reservoirs (Ahmed & Meehan, 2016). Model of Akkutlu and Fathi (2012) includes dual porosity continua of matrix/fracture system, where the matrix is comprised of both organic and inorganic pores. Sakhaee-Pour and Bryant (2012) developed model including spatial characterization and geometry of porous media using slip flow assumption, represented by Maxwell theory. Naraghi and Javadpour (2015) presented a model developed by stochastically characterizing organic and inorganic pores. This model can distinguish different pore systems in organic and inorganic matters. Sheng et al. (2018) presented gas transport models in shale considering the change in pore size due to the applied stress. They provided a compression coefficient to characterize the influence of stress sensitivity on key parameters for gas transport. In spite of numerous researches in this subject, there is no consensus of fluid flow in nanoscale flow of shale reservoirs so that continuous research and rigorous verification are required.

MOLECULAR DIFFUSION

Because of the low permeability of shale reservoirs, the effect of molecular diffusion is significant compared with conventional reservoirs. Notably, several kinds of literature (Kim et al., 2017; Sheng, 2015; Wang, Luo, Er, & Huang, 2010; Yu, Lashgari, Wu, & Sepehrnoori, 2015) presented that molecular diffusion should be considered during CO_2 injection in shale reservoirs, but it is still poorly understood. Molecular diffusion driven by differences in concentration gradient is typically modeled by Fick's law (1855). Fick's law relates the diffusive flux to the concentration under the assumption of steady state. It postulates that the flux goes from regions of high concentration to areas of low concentration with a magnitude that is proportional to the concentration gradient. The generalized form of the Fick's first law is

$$J_d = -D\nabla C, \tag{3.71}$$

where J_d is the molecular diffusion flux, D the diffusion coefficient, and C the concentration. If concentration gradients exist, gas diffusion will cause the concentration to equilibrate gradually. According to Freeman et al. (2011), when Knudsen diffusion causes gas species to be separated, molecular diffusion will counteract any fractionating effects. Consequently, it is necessary to include the gas diffusion term in shale reservoir models.

There are several methods to obtain molecular diffusion coefficients for calculations. Based on various experimental data, Wilke and Chang (1955) showed that the diffusion factor, $\frac{D_i\mu}{T}$, is correlated with the solute molal volume and solvent molecular weight. From these results, they obtained the following equation:

$$D_i = \frac{7.4 \times 10^{-8}\left(M'_i\right)^{0.5}T}{\mu v_{bi}^{0.6}}, \quad (3.72)$$

where

$$M'_i = \frac{\sum\limits_{j \neq i} \gamma_j M_j}{1 - \gamma_i}, \quad (3.73)$$

$$v_{bi} = 0.285 v_c^{1.048}, \quad (3.74)$$

D_i is the diffusion coefficient of the component i in the mixture, M'_i the solvent molecular weight, v_{bi} the partial molal volume of component i at the boiling point, γ_i the mole fraction of component i, M_j the molecular weight of the component j, and v_c the critical volume.

Fitting experimental results, Sigmund (1976a, 1976b) also obtained a generalized correlation for predicting binary molecular diffusion coefficients. The correlation is as follows:

$$D_{ij} = \frac{\rho^0 D_{ij}^0}{\rho}\left(0.99589 + 0.096016\rho_r - 0.22035\rho_r^2 + 0.032874\rho_r^3\right), \quad (3.75)$$

where

$$\rho_r = \rho\left(\frac{\sum\limits_{i=1}^{n_c} \gamma_i v_{ci}^{\frac{5}{3}}}{\sum\limits_{i=1}^{n_c} \gamma_i v_{ci}^{\frac{2}{3}}}\right), \quad (3.76)$$

$$\rho^0 D_{ij}^0 = \frac{0.0018583 T^{0.5}}{\sigma_{ij}^2 \Omega_{ij} R}\left(\frac{1}{M_i} + \frac{1}{M_j}\right)^{0.5}, \quad (3.77)$$

D_{ij} is the binary diffusion coefficient between the component i and j in the mixture, $\rho^0 D_{ij}^0$ is the zero-pressure limit of the density-diffusivity product, ρ is the molar density of the diffusing mixture, and ρ_r is the reduced density. The values of $\rho^0 D_{ij}^0$ were calculated from Chapman-Enskog dilute gas theory (Hirschfelder, Curtiss, Bird, & University of Wisconsin. Theoretical Chemistry Laboratory, 1954) using the Stiel-Thodos

correlation (Stiel and Thodos, 1962) for the estimation of the molecular parameters. The dimensionless collision integral Ω_{ij} of the Lennard-Jones potential and the collision diameter σ_{ij} are related to the component critical properties through the following equations (Reid, 1977):

$$\Omega_{ij} = 1.06306\left(T_{ij}^*\right)^{-0.1561} + 0.193e^{-0.47635T_{ij}^*} + 1.03587e^{-1.52996T_{ij}^*} + 1.76474e^{-3.89411T_{ij}^*}, \quad (3.78)$$

$$\sigma_{ij} = \frac{\sigma_i + \sigma_j}{2}, \quad (3.79)$$

$$\sigma_i = (2.3551 - 0.087\omega_i)\left(\frac{T_{ci}}{p_{ci}}\right)^{\frac{1}{3}}, \quad (3.80)$$

$$T_{ij}^* = \frac{k_B}{\varepsilon_{ij}}T, \quad (3.81)$$

$$\varepsilon_{ij} = \sqrt{\varepsilon_i \varepsilon_j}, \quad (3.82)$$

$$\varepsilon_i = k_B(0.7915 + 0.1963\omega_i)T_{ci}, \quad (3.83)$$

where p_c is the critical pressure, T_c the critical temperature, and ε the Lennard-Jones energy. Using the binary diffusion coefficient, the diffusion coefficient of the component i can be calculated as follows:

$$D_i = \frac{1 - \gamma_i}{\sum\limits_{j \neq i} \gamma_j D_{ij}^{-1}}. \quad (3.84)$$

Meanwhile, depending on Webb and Pruess (2003), the application of Fick's law into a convective flow equation is inappropriate to model low permeability porous media. They described that the use of the advective-diffusive flow model, known as the extended Fick's law model, should be restricted to media with the permeability of 1000 md or larger. Due to the low permeability of shale matrix (1−100 nd), the advective-diffusive flow model cannot be used in shale reservoirs. To consider effects of molecular diffusion with viscous and Knudsen flow in porous media, dusty-gas model could be applied (Freeman et al., 2011; Krishna, 1993; Li, Zhang, & Liu, 2017; Mason & Malinauskas, 1983; Yao et al., 2013; Zeng et al., 2017). In the dusty-gas model, apparent permeability is presented in the form of the Klinkenberg effect for one component in shale gas reservoirs with the viscous flow and Knudsen diffusion. The dusty-gas model is derived from the full Chapman−Enskog kinetic theory of gases (Sumner, 1999) as shown below:

$$\sum_{j=1,j \neq i}^{n} \frac{\gamma_i J_j - \gamma_j J_i}{D_{ij}^e} - \frac{J_i}{D_{K,i}} = \frac{p}{RT}\nabla\gamma_i + \left(1 + \frac{k_D p}{\mu D_{K,i}}\right)\frac{\gamma_i \nabla p}{RT}, \quad (3.85)$$

where D_{ij}^e is the effective gas diffusivity of species i in species j and n is the number of components present in the system. The dusty-gas model is a system of equations, with the number of gas species being accounted for. If only one species is present ($n = 1$, $\gamma_1 = 1$), this equation reduces to:

$$J_i = -\left(D_{K,i} + \frac{k_D p}{\mu}\right) \frac{\nabla p}{RT}.$$ (3.86)

For a binary system, equations can be solved simultaneously though systems of more than two components cannot be solved explicitly. Using Knudsen diffusion and molecular diffusion coefficients, the composition of the flux flowing through the porous medium is calculated by equations of the dusty-gas model.

GEOMECHANICS

Production performances of shale plays depend strongly on the existence of a dense and conductive network of fractures (Cho, Apaydin, & Ozkan, 2013). The conductivity of fracture network is sensitive to the change of stress and strain during production because of stress corrosion affecting the proppant strength, crushing, and embedment into the formation (Ghosh, Rai, Sondergeld, & Larese, 2014). Therefore, geomechanic effects during production must be included to consider a variation of conductivity in shale reservoirs. In previous studies, simple pressure-dependent properties were presented to recognize the change of conductivity, and they showed inaccurate results (Cho et al., 2013; Pedrosa, 1986; Raghavan & Chin, 2004). In the following section, therefore, stress-dependent properties with the coupling of fluid flow and geomechanic calculations were introduced.

The two critical elements of geomechanics are the internal resistance of a solid object, which acts to balance the effects of imposing external forces, represented by a term called stress, and the shape change and deformation of the solid object in response to external forces, denoted by strain (Aadnoy & Looyeh, 2011). In general, the stress σ across any horizontal surface can be expressed and calculated directly with the magnitude of those forces, F and cross-sectional area, A:

$$\sigma = \frac{F}{A}.$$ (3.87)

Stress is independent of the size and shape of the body. When a body is subjected to loading, it will undergo deformation. Deformation is typically quantified regarding the original dimension, and it is represented by strain. The strain is, therefore, defined as deformation divided by the nondeformed dimension, l:

$$\varepsilon = \frac{\Delta l}{l},$$ (3.88)

where ε is the strain, Δl the deformed dimension, and l the original nondeformed dimension.

The degree of a material strain depends on the magnitude of imposed stress. For most materials that are stressed at a low level, the strain is proportional to stress with a simple linear relationship (Fig. 3.13) as shown in the following equation:

$$\sigma = E\varepsilon.$$ (3.89)

The linear relationship is known as Hooke's law for elastic material and E is Young's modulus or elastic modulus. E is defined as the ratio of tension to an extension in a rod which is under axial tension and which is unrestricted laterally. The higher the modulus, the more stress is needed to create the same amount of strain.

In general, a material tends to expand in directions perpendicular to the direction of compression. Conversely, if the material is stretched, it usually tends to contract in the directions transverse to the direction of stretching. Poisson's ratio is a measure of this phenomenon. Poisson's ratio is defined as the ratio of lateral contraction to longitudinal extension. The Poisson's ratio, ν, is expressed as:

$$\nu = \frac{\varepsilon_y}{\varepsilon_x},$$ (3.90)

where ε_x is an axial strain, and ε_y is a transverse strain. Most materials have Poisson's ratio values ranging between 0.0 and 0.5. A perfectly incompressible material deformed elastically at small strains would have a Poisson's ratio of exactly 0.5. Young's modulus and Poisson's ratio are significant basic material properties for following calculations of the geomechanic model.

Basic equations for a deformable porous medium are presented by Tran, Nghiem, and Buchanan (2005a). They assumed homogeneous, isotropic, and symmetric rock material and very small strain compared to unity. Force equilibrium equation in three spatial scales is shown below:

$$\nabla \cdot \boldsymbol{\sigma} - \rho_r \mathbf{B} = 0,$$ (3.91)

where $\boldsymbol{\sigma}$ is the stress tensor and $\mathbf{B}$ is the force per unit mass that accounts for gravity. Let $\mathbf{u}$ be the displacement vector that connects the position of a particular particle in the reference configuration to its new position in a deformed configuration. The gradient of the displacement vector $\mathbf{u}$ can be decomposed as:

$$\nabla \mathbf{u} = \frac{1}{2}\left[\nabla \mathbf{u} + (\nabla \mathbf{u})^T\right] + \frac{1}{2}\left[\nabla \mathbf{u} - (\nabla \mathbf{u})^T\right],$$ (3.92)

where the superscript T denotes the matrix transpose. On the right-hand side of Eq. (3.92), the first term is a symmetric matrix equivalent to the strain tensor, $\boldsymbol{\varepsilon}$,

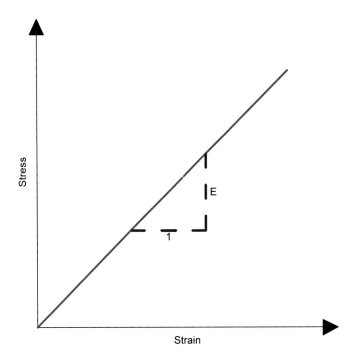

FIG. 3.13 Stress-strain relationship with linear elastic deformation.

which is a result of changing length or shape within a body. The second term is a skew-symmetric matrix equivalent to the rotation tensor, $\mathbf{R}$, which is a result of moving a rigid body as given by (Davis & Selvadurai, 1996):

$$\boldsymbol{\varepsilon} = \frac{1}{2}\left[\nabla\mathbf{u} + (\nabla\mathbf{u})^T\right], \tag{3.93}$$

$$\mathbf{R} = \frac{1}{2}\left[\nabla\mathbf{u} - (\nabla\mathbf{u})^T\right]. \tag{3.94}$$

The stress $\boldsymbol{\sigma}$ expressed in Eq. (3.91) is the total stress tensor. However, in a porous medium, effective stress, $\boldsymbol{\sigma}'$, which is part of the total stress, affects the strength of solid rock grains (Terzaghi, 1943). The total stress and effective stress are related through the formula given by (Biot & Willis, 1957):

$$\boldsymbol{\sigma} = \boldsymbol{\sigma}' + \alpha p\mathbf{I}, \tag{3.95}$$

where $\mathbf{I}$ is the identity matrix and parameter, α is the Biot's constant, which has a value between the porosity and unity. In unconsolidated or weak rock, α is closed to unity. The relationship among total stress, effective stress, and pore pressure is illustrated in Fig. 3.14.

The constitutive relationship between stress, strain, and the temperature is key to geomechanic processes. For a thermoporous medium, a linear constitutive relation can be written for a single dimension as:

$$\sigma' = E(\varepsilon - \beta_r \Delta T), \tag{3.96}$$

where β_r is the linear thermal expansion coefficient of the solid rock. For multiple dimensions, the general constitutive relation is:

$$\boldsymbol{\sigma}' = \mathbf{C} : \boldsymbol{\varepsilon} - \eta \Delta T \mathbf{I}, \tag{3.97}$$

where $\mathbf{C}$ is the tangential stiffness tensor (equivalent to Young's modulus in a one-dimensional linear case) and $\eta = \frac{E\beta_r}{(1-2\nu)}$ for 3D and plane strain and $\eta = \frac{E\beta_r}{(1-\nu)}$ for plane stress. Substituting Eqs. (3.93), (3.95) and (3.97) into force equilibrium equation leads to the displacement equation:

$$\nabla \cdot \left[\mathbf{C} : \frac{1}{2}\left(\nabla\mathbf{u} + (\nabla\mathbf{u})^T\right)\right] + \nabla \cdot \left[(\alpha p - \eta \Delta T)\mathbf{I}\right] = \rho_r \mathbf{B}. \tag{3.98}$$

To model the accurate shale reservoir production considering geomechanic deformation, two sets of equations for fluid flow in porous media and solid deformation should be coupled. The fluid flow

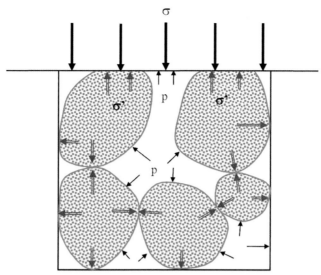

FIG. 3.14 Relationship among total stress, effective stress, and pore pressure. (Credit: Tran, D., Nghiem, L., & Buchanan, L. (2005a). An overview of iterative coupling between geomechanical deformation and reservoir flow. *Paper presented at the SPE international thermal operations and heavy oil symposium, Calgary, Alberta, Canada.* https://doi.org/10.2118/97879-MS.)

equation including conservation of mass, Darcy's law and equation of state for fluid is shown as follows:

$$\frac{\partial}{\partial t}(\phi\rho) - \nabla \cdot \left[\rho \frac{\mathbf{k}}{\mu}(\nabla p - \rho_f \mathbf{b}) \right] = Q_f, \qquad (3.99)$$

where $\mathbf{k}$ it the absolute permeability tensor, $\mathbf{b}$ the body force per unit mass of fluid, and Q_f the flow rate of the fluid at the source or sink location.

Several methods for coupling geomechanics to fluid flow in the reservoir have been proposed in the literature: fully coupled, iterative coupled and explicit coupled methods (Samier, Onaisi, & Fontaine, 2003; Tran, Settari, & Nghiem, 2004; Tran et al., 2005a; Tran, Nghiem, & Buchanan, 2005b). The fully coupled approach is the tightest coupling because reservoir pressure and temperature and deformation are solved simultaneously. The solution is reliable and can be used as a benchmark for other coupling approaches. However, this coupling is not widely used for large-scale simulation and nonlinear problems. The solution process is very time-consuming as it requires the simultaneous solution of the flow variables (pressure, saturation, composition, and temperature) and the geomechanics variables (displacements, stresses, and strains). The iterative coupled approach is less tight than the full coupling method. The geomechanics calculations are not performed at the same time as the reservoir flow calculations but one step behind. In this

coupling, the information computed in the reservoir simulator and the geomechanics module is exchanged back and forth. Therefore, the reservoir flow is affected by the geomechanic responses. The explicit coupled approach is considered as a special case of the iterative coupled approach. The information from a reservoir simulator is sent to a geomechanics module, but the calculations in the geomechanics module are not fed back to the reservoir simulator. In this case, the reservoir flow is not affected by the geomechanic responses calculated by the geomechanics module.

Due to accuracy and efficiency of computational calculation, the iterative coupled method has been widely used. The pressure p for a given time step is obtained from Eq. (3.99) and then, p is used in displacement equation to obtain the displacement vector $\mathbf{u}$. After the displacement $\mathbf{u}$ is determined, the strain tensor and stress tensor can be computed. A link between geomechanics and flow can be represented by the following porosity equation which is function of pressure, temperature, and total mean stress (Tran, Buchanan, and Nghiem, 2010; Tran, Nghiem, and Buchanan, 2009; Tran, Settari, & Nghiem, 2002):

$$\phi_{n+1} = \phi_n + (c_0 + c_2 a_1)(p - p^n) + (c_1 + c_2 a_2)(T - T^n), \qquad (3.100)$$

where

$$c_0 = \frac{1}{V_{b,i}} \left(\frac{dV_p}{dp} + V_b \alpha c_b \frac{d\sigma_m}{dp} - V_p \beta \frac{dT}{dp} \right), \qquad (3.101)$$

$$c_1 = \frac{V_p}{V_{b,i}} \beta_r, \qquad (3.102)$$

$$c_2 = -\frac{V_b}{V_{b,i}} \alpha c_b, \qquad (3.103)$$

$$a_1 = \Gamma \left[\frac{2}{9} \frac{E}{(1-\nu)} \alpha c_b \right], \qquad (3.104)$$

$$a_2 - \Gamma \left[\frac{2}{9} \frac{E}{(1-\nu)} \beta \right], \qquad (3.105)$$

V_b is the bulk volume, V_p the pore volume, c_b the bulk compressibility, σ_m the mean total stress, and β the volumetric thermal expansion coefficient. The subscript i indicates the initial state. ϕ is used to Eq. (3.99) calibrated at each time step during the coupling iteration loop between fluid flow and deformation equations.

Reservoir fluid flow model is coupled with a geomechanic model, and then, stress-dependent correlations are used to consider porosity and permeability variation. Dong et al. (2010) presented experimental results measuring porosity and permeability concerning effective confining pressure. Fig. 3.15 shows that results of curve fitting on measured porosity and permeability of the sandstone and shale cores with exponential and power law correlations. They used exponential and power law correlations to match the experimental data as follows:

$$\phi = \phi_i e^{-a(\sigma' - \sigma_i')}, \qquad (3.106)$$

$$k = k_i e^{-b(\sigma' - \sigma_i')}, \qquad (3.107)$$

$$\phi = \phi_i \left(\frac{\sigma'}{\sigma_i'} \right)^{-c}, \qquad (3.108)$$

$$k = k_i \left(\frac{\sigma'}{\sigma_i'} \right)^{-d}, \qquad (3.109)$$

where a, b, c, and d are experimental coefficients. These parameters were determined using curve fitting techniques on measured porosity and permeability of the shale cores. Porosity and permeability are generated by the geomechanics simulator and passed to the fluid flow simulator.

PHASE BEHAVIOR IN NANOPORES

Shale matrix is characterized by very low permeability and small pore size, typically in nano-Darcy (nd) and nanometer (nm) scales, respectively (Ambrose, Hartman, & Yucel Akkutlu, 2011; Curtis, Joseph Ambrose, Carl, & Sondergeld, 2010; Nelson, 2009; Sigal, 2015; Sondergeld, Ambrose, Rai, & Moncrieff, 2010). Because of the tiny pore sizes, the capillary pressure is significantly high, which changes the phase behavior of the fluid in the confining space of the nanopores. This phenomenon, as the result of the nanoscale pore and high capillary pressure, is called the capillary condensation or confinement effect that is the condensation of the vapor inside nanoscale pores. In spite of the effects of capillary condensation for shale reservoirs, most current researches of reservoir simulation in shale consider only adsorbed and free gas on the pore walls and in the pore bodies. Chen, Mehmani, Li, Georgi, and Jin (2013) estimated in principle that accounting for capillary condensation could increase reserve estimations by up to three to six times in shale reservoirs. Because the phase behavior of fluids is different in nanopores, more reliable models are needed in shale reservoirs.

In a confined space, physical behavior of fluids is different from that in the bulk (Barsotti, Tan, Saraji, Piri, & Chen, 2016). In nanoporous media, the effects of molecular size and mean free path should be considered to analyze the deviation from the conventional fluid flow. At this scale, distances between molecules are short so that intermolecular forces are substantial. Subsequently, phase behavior in nanopores is affected by fluid-pore wall interactions as well as fluid-fluid interactions as it is in the unconfined pores. Depending on Barsotti et al. (2016), capillary and adsorptive forces alter phase boundaries (Alharthy, Nguyen, Teklu, Kazemi, & Graves, 2013; Casanova et al., 2008; Du & Chu, 2012; Evans et al., 1986; Gelb, Gubbins, Radhakrishnan, & Sliwinska-Bartkowiak, 1999; Gubbins, Long, & Sliwinska-Bartkowiak, 2014; Jin, Ma, & Ahmad, 2013; Nojabaei, Johns, & Chu, 2013; Russo, Carrott, & Carrott, 2012; Teklu et al., 2014; Thommes & Cychosz, 2014; Thommes & Findenegg, 1994; Yun, Duren, Keil, & Seaton, 2002; Zhang, Civan, Devegowda, & Sigal, 2013), phase compositions (Gelb et al., 1999; Gubbins et al., 2014; Radhakrishnan, Gubbins, & Sliwinska-Bartkowiak, 2002; Zhang et al., 2013), interfacial tensions (Du & Chu, 2012), densities (Cole, Ok, Striolo, & Anh, 2013; Gelb et al., 1999; Jin et al., 2013; Keller & Staudt, 2005; Nojabaei et al., 2013; Thommes & Findenegg, 1994), viscosities (Alharthy et al., 2013; Du & Chu, 2012), and saturation pressures (Chen et al., 2013; Evans et al., 1986; Mitropoulos, 2008; Naumov, Valiullin, Monson, & Karger, 2008; Nojabaei et al., 2013). Understanding of the physical behavior variation in confined fluids is significant to improve the insights of various disciplines: catalysis (Gelb et al., 1999; Thommes & Cychosz, 2014; Ye, Zhou, Yuan,

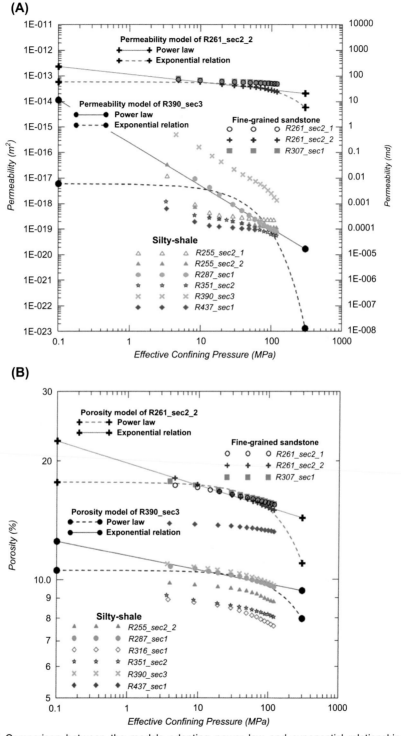

FIG. 3.15 Comparison between the models adopting power law and exponential relationships for **(A)** porosity and **(B)** permeability fitting. (Credit: Dong, J. J., Hsu, J. Y., Wu, W. J., Shimamoto, T., Hung, J. H., Yeh, E. C., et al. (2010). Stress-dependence of the permeability and porosity of sandstone and shale from ICDP Hole-A International Journal of Rock Mechanics and Mining Sciences, 47(7), 1141–1157. https://doi. org/10.1016/j.ijrmms.2010.06.019.)

Ye, & Coppens, 2016), chemistry (Long et al., 2013), geochemistry (Cole et al., 2013), geophysics (Gelb et al., 1999), nanomaterials (Gelb et al., 1999), battery design (Thommes & Cychosz, 2014), carbon dioxide sequestration (Belmabkhout, Serna-Guerrero, & Sayari, 2009; Yun et al., 2002), drug delivery (Thommes & Cychosz, 2014), enhanced coalbed methane recovery (Gor et al., 2013), lubrication and adhesion (Gelb et al., 1999), materials characterization (Kruk, Jaroniec, & Sayari, 1997; Kruk, Jaroniec, & Sayari, 1999; Kruk, Jaroniec, Ko, & Ryoo, 2000; Ravikovitch & Neimark, 2001; Tanchoux, Trens, Maldonado, Di Renzo, & Fajula, 2004), micro/nanoelectromechanical system design, pollution control (Gelb et al., 1999; Mower, 2005; Shim, Lee, & Moon, 2006; Yun et al., 2002), and separation (Thommes & Cychosz, 2014), as well as fluid flow analysis in shale and tight formations.

In unconfined space, gas condenses into a liquid if the gas pressure is equal to its dew point pressure or higher. However, gas condenses into liquid at a pressure lower than its dew point pressure in confined spaces such as nanopores (Chen et al., 2012; Gelb et al., 1999; Li, Mehmani, Chen, Georgi, & Jin, 2013; de Keizer et al., 1991). In porous media, gas in a pore with radius r condenses to a liquid if the pressure satisfies the Kelvin equation as follows (Thomson, 1872):

$$p \geq p_d \exp\left(-\frac{2\gamma V_L \cos \theta_c}{rRT}\right), \qquad (3.110)$$

In this equation, p_d is the dew point pressure, γ the interfacial tension (IFT), V_L the liquid molar volume, and θ_c the contact angle. From Eq. (3.110), the following equation for the critical pore radius can be derived:

$$r_c = -\frac{2\gamma V_L \cos \theta_c}{RT \ln\left(\frac{p}{p_d}\right)}. \qquad (3.111)$$

In the pores with a smaller size of radius than the critical radius, gas condenses to the liquid while the larger pores remain as gas. The presence of a condensed phase would reduce the permeability in the reservoir as it blocks the gas flow. If a pore has diameter variation, the two phases of gas and oil can exist along the pore. Confinement by nanopores may introduce different conditions when two phases coexist. This greatly complicates the transport of fluid in nanopores of shale reservoirs compared to that in conventional reservoirs. The effects of accounting for capillary condensation on the analysis of shale reservoir production are still unknown so that reliable models are needed to obtain more accurate estimates.

Singh, Sinha, Deo, and Singh (2009) and Travalloni, Castier, Tavares, and Sandler (2010) have quantified the changes of critical pressure in confined pores. Hamada, Koga, and Tanaka (2007) simulated the thermodynamic behavior of confined Lennard-Jones (LJ) particles in the slit and cylindrical pores using the grand canonical Monte Carlo simulation. They concluded that thermodynamic properties and fluid phase behavior between the fluid and pore surface are altered in the nanoporous media. Zarragoicoechea and Kuz (2004) modeled the reduction in critical temperature by applying the van der Waals model, and they reported a good match between their model and the experimental data obtained by Morishige and Nobuoka (1997). They presented the relative critical temperature shift correlating quadratically with the ratio of the LJ collision diameter to the pore throat radius, $\frac{\sigma_{LJ}}{r_p}$ (Eq. 3.112) where σ_{LJ}, the collision diameter, is calculated based on the bulk critical properties using Eq. (3.114) (Bird et al. 2007). The critical pressure shift shows a similar dependence on $\frac{\sigma_{LJ}}{r_p}$ as the critical temperature shift based on the van der Waals theory (Eq. 3.113).

$$\Delta T_c^* = \frac{T_{cb} - T_{cp}}{T_{cb}} = 0.9409 \frac{\sigma_{LJ}}{r_p} - 0.2415 \left(\frac{\sigma_{LJ}}{r_p}\right)^2, \qquad (3.112)$$

$$\Delta p_c^* = \frac{p_{cb} - p_{cp}}{p_{cb}} = 0.9409 \frac{\sigma_{LJ}}{r_p} - 0.2415 \left(\frac{\sigma_{LJ}}{r_p}\right)^2, and \qquad (3.113)$$

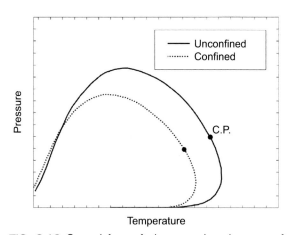

FIG. 3.16 General form of phase envelope in cases of confined and unconfined pores. (Credit: Barsotti, E., Tan, S. P., Saraji, S., Piri, M., & Chen, J. H. (2016). A review on capillary condensation in nanoporous media: Implications for hydrocarbon recovery from tight reservoirs. *Fuel, 184,* 344–361. https://doi.org/10.1016/j.fuel.2016.06.123.)

$$\sigma_{LJ} = 0.244\sqrt[3]{\frac{T_{cb}}{p_{cb}}}, \quad\quad (3.114)$$

where ΔT_c^* is the relative critical temperature shift, T_{cb} is the bulk critical temperature, T_{cp} is the pore critical temperature, r_p is the pore throat radius, Δp_c^* is the relative critical pressure shift, p_{cb} is the bulk critical pressure, and p_{cp} is the pore critical pressure. Fig. 3.16 shows phase behavior shift depending on confinement effects.

REFERENCES

Aadnoy, B., & Looyeh, R. (2011). *Petroleum rock mechanics: Drilling operations and well design.* Elsevier Science.

Adamson, A. W., & Gast, A. P. (1997). *Physical chemistry of surfaces.* Wiley.

Agrawal, A., & Prabhu, S. V. (2008). Survey on measurement of tangential momentum accommodation coefficient. *Journal of Vacuum Science & Technology A, 26*(4), 634–645. https://doi.org/10.1116/1.2943641.

Ahmed, U., & Meehan, D. N. (2016). *Unconventional oil and gas resources: Exploitation and development.* CRC Press.

Akkutlu, I. Y., Efendiev, Y., & Savatorova, V. (2015). Multi-scale Asymptotic analysis of gas transport in shale matrix. *Transport in Porous Media, 107*(1), 235–260. https://doi.org/10.1007/s11242-014-0435-z.

Akkutlu, I. Y., & Fathi, E. (2012). Multiscale gas transport in shales with local kerogen heterogeneities. *SPE Journal, 17*(4), 1002–1011. https://doi.org/10.2118/146422-Pa.

Alharthy, N. S., Nguyen, T., Teklu, T., Kazemi, H., & Graves, R. (2013). Multiphase compositional modeling in small-scale pores of unconventional shale reservoirs. In *Paper presented at the SPE annual technical conference and exhibition, New Orleans, Louisiana, USA.* https://doi.org/10.2118/166306-MS.

Ambrose, R. J., Hartman, R. C., & Yucel Akkutlu, I. (2011). Multi-component sorbed phase considerations for shale gas-in-place calculations. In *Paper presented at the SPE production and operations symposium, Oklahoma city, Oklahoma, USA.* https://doi.org/10.2118/141416-MS.

Arkilic, E. B. (1997). *Measurement of the mass flow and tangential momentum accommodation coefficient in silicon micromachined channels* (Ph.D.). Cambridge: Fluid Dynamics Research Laboratory Department of Aeronautics and Astronautics Massachusetts Institue of Technology.

Azom, P. N., & Javadpour, F. (2012). Dual-continuum modeling of shale and tight gas reservoirs. In *Paper presented at the SPE annual technical conference and exhibition, San Antonio, Texas, USA.* https://doi.org/10.2118/159584-MS.

Barsotti, E., Tan, S. P., Saraji, S., Piri, M., & Chen, J. H. (2016). A review on capillary condensation in nanoporous media: Implications for hydrocarbon recovery from tight reservoirs. *Fuel, 184*, 344–361. https://doi.org/10.1016/j.fuel.2016.06.123.

Bear, J. (1988). *Dynamics of fluids in porous media.* New York (N.Y.): Dover.

Belmabkhout, Y., Serna-Guerrero, R., & Sayari, A. (2009). Adsorption of CO_2 from dry gases on MCM-41 silica at ambient temperature and high pressure. 1: Pure CO_2 adsorption. *Chemical Engineering Science, 64*(17), 3721–3728. https://doi.org/10.1016/j.ces.2009.03.017.

Beskok, A., & Karniadakis, G. E. (1999). A model for flows in channels, pipes, and ducts at micro and nano scales. *Microscale Thermophysical Engineering, 3*(1), 43–77.

Biot, M. A., & Willis, D. G. (1957). The elastic coefficients of the theory of consolidation. *Journal of Applied Mechanics, 24*, 594–601. https://doi.org/citeulike-article-id:9228504.

Bird, R. B., Stewart, W. E., & Lightfoot, E. N. (2007). *Transport phenomena.* Wiley.

Brillouin, M. (1907). *Leçons sur la viscosité des liquides et des gaz: Pt. Viscosité des gaz. Caractéres généraux des théories moléculaires.* Gauthier-Villars.

Brown, G. P., Dinardo, A., Cheng, G. K., & Sherwood, T. K. (1946). The flow of gases in pipes at low pressures. *Journal of Applied Physics, 17*(10), 802–813. https://doi.org/10.1063/1.1707647.

Brunauer, S., Emmett, P. H., & Teller, E. (1938). Adsorption of gases in multimolecular layers. *Journal of the American Chemical Society, 60*, 309–319. https://doi.org/10.1021/ja01269a023.

Busch, A., Alles, S., Gensterblum, Y., Prinz, D., Dewhurst, D. N., Raven, M. D., et al. (2008). Carbon dioxide storage potential of shales. *International Journal of Greenhouse Gas Control, 2*(3), 297–308. https://doi.org/10.1016/j.ijggc.2008.03.003.

Cao, C., Li, T., Shi, J., Zhang, L., Fu, S., Wang, B., et al. (2016). A new approach for measuring the permeability of shale featuring adsorption and ultra-low permeability. *Journal of Natural Gas Science and Engineering, 30*, 548–556. https://doi.org/10.1016/j.jngse.2016.02.015.

Casanova, F., Chiang, C. E., Li, C. P., Roshchin, I. V., Ruminski, A. M., Sailor, M. J., et al. (2008). Effect of surface interactions on the hysteresis of capillary condensation in nanopores. *Epl, 81*(2). https://doi.org/10.1209/0295-5075/81/26003.

Chen, J.-H., Mehmani, A., Li, B., Georgi, D., & Jin, G. (2013). Estimation of total hydrocarbon in the presence of capillary condensation for unconventional shale reservoirs. In *Paper presented at the SPE Middle East oil and gas show and conference, Manama, Bahrain.* https://doi.org/10.2118/164468-MS.

Chen, J.-H., Zhang, J., Jin, G., Quinn, T., Frost, E., & Chen, J. (2012). Capillary condensation and NMR relaxation time in unconventional shale hydrocarbon resources. In *Paper presented at the SPWLA 53rd annual logging symposium, Cartagena, Colombia.*

Cho, Y., Apaydin, O. G., & Ozkan, E. (2013). Pressure-dependent natural-fracture permeability in shale and its effect on shale-gas well production. *SPE Reservoir Evaluation & Engineering, 16*(2), 216–228. https://doi.org/10.2118/159801-Pa.

Cipolla, C. L., Lolon, E. P., Erdle, J. C., & Rubin, B. (2010). Reservoir modeling in shale-gas reservoirs. *SPE Reservoir Evaluation & Engineering, 13*(4), 638–653. https://doi.org/10.2118/125530-Pa.

Civan, F. (2007). *Reservoir formation damage: Fundamentals, modeling, assessment, and mitigation.* Gulf Professional Pub.

Civan, F. (2010). Effective correlation of apparent gas permeability in tight porous media. *Transport in Porous Media, 82*(2), 375–384. https://doi.org/10.1007/s11242-009-9432-z.

Civan, F., Devegowda, D., & Sigal, R. F. (2013). Critical evaluation and improvement of methods for determination of matrix permeability of shale. In *Paper presented at the SPE annual technical conference and exhibition, New Orleans, Louisiana, USA.* https://doi.org/10.2118/166473-MS.

Civan, F., Rai, C. S., & Sondergeld, C. H. (2011). Shale-gas permeability and diffusivity inferred by improved formulation of relevant retention and transport mechanisms. *Transport in Porous Media, 86*(3), 925–944. https://doi.org/10.1007/s11242-010-9665-x.

Cole, D. R., Ok, S., Striolo, A., & Anh, P. (2013). Hydrocarbon behavior at nanoscale interfaces. *Carbon in Earth, 75*, 495–545. https://doi.org/10.2138/rmg.2013.75.16.

Coles, M. E., & Hartman, K. J. (1998). Non-Darcy measurements in dry core and the effect of immobile liquid. In *Paper presented at the SPE gas technology symposium, Calgary, Alberta, Canada.* https://doi.org/10.2118/39977-MS.

Cooke, C. E. (September 1973). Conductivity of fracture proppants in multiple layers. *Journal of Petroleum Technology, 25*, 1101–1107. https://doi.org/10.2118/4117-Pa.

Cooper, J. W., Wang, X. L., & Mohanty, K. K. (1999). Non-Darcy-Flow studies in anisotropic porous media. *SPE Journal, 4*(4), 334–341. https://doi.org/10.2118/57755-Pa.

Coppens, M. O. (1999). The effect of fractal surface roughness on diffusion and reaction in porous catalysts — from fundamentals to practical applications. *Catalysis Today, 53*(2), 225–243. https://doi.org/10.1016/S0920-5861(99)00118-2.

Coppens, M. O., & Dammers, A. J. (2006). Effects of heterogeneity on diffusion in nanopores - from inorganic materials to protein crystals and ion channels. *Fluid Phase Equilibria, 241*(1–2), 308–316. https://doi.org/10.1016/j.fluid.2005.12.039.

Cornell, D., & Katz, D. L. (1953). Flow of gases through consolidated porous media. *Industrial and Engineering Chemistry, 45*(10), 2145–2152. https://doi.org/10.1021/ie50526a021.

Curtis, M. E., Joseph Ambrose, R., Carl, H., & Sondergeld. (2010). Structural characterization of gas shales on the micro- and nano-scales. In *Paper presented at the Canadian unconventional resources and international petroleum conference, Calgary, Alberta, Canada.* https://doi.org/10.2118/137693-MS.

Darabi, H., Ettehad, A., Javadpour, F., & Sepehrnoori, K. (2012). Gas flow in ultra-tight shale strata. *Journal of Fluid Mechanics, 710*, 641–658. https://doi.org/10.1017/jfm.2012.424.

Darcy, H. (1856). *Les fontaines publiques de la ville de Dijon: Exposition et application des principes a suivre et des formules a employer dans les questions de distribution d'eau; ouvrage* terminé par un appendice relatif aux fournitures d'eau de plusieurs villes au filtrage des eaux et a la fabrication des tuyaux de fonte. de plomb, de tole et de bitume: Victor Dalmont, Libraire des Corps imperiaux des ponts et chaussées et des mines.

Davis, R. O., & Selvadurai, A. P. S. (1996). *Elasticity and geomechanics.* Cambridge University Press.

Dong, J. J., Hsu, J. Y., Wu, W. J., Shimamoto, T., Hung, J. H., Yeh, E. C., et al. (2010). Stress-dependence of the permeability and porosity of sandstone and shale from TCDP Hole-A. *International Journal of Rock Mechanics and Mining Sciences, 47*(7), 1141–1157. https://doi.org/10.1016/j.ijrmms.2010.06.019.

Donohue, M. D., & Aranovich, G. L. (1998). Classification of Gibbs adsorption isotherms. *Advances in Colloid and Interface Science, 76*, 137–152. https://doi.org/10.1016/S0001-8686(98)00044-X.

Du, L., & Chu, L. (2012). Understanding anomalous phase behavior in unconventional oil reservoirs. In *Paper presented at the SPE Canadian unconventional resources conference, Calgary, Alberta, Canada.* https://doi.org/10.2118/161830-MS.

Ergun, S. (1952). Fluid flow through packed columns. *Chemical Engineering Progress, 48*(2), 89–94.

Ergun, S., & Orning, A. A. (1949). Fluid flow through randomly packed columns and fluidized beds. *Industrial and Engineering Chemistry, 41*(6), 1179–1184. https://doi.org/10.1021/ie50474a011.

Evans, R. D., & Civan, F. (1994). *Characterization of non-Darcy multiphase flow in petroleum bearing formation.* Final report. United States.

Evans, R. D., Hudson, C. S., & Greenlee, J. E. (1987). The effect of an immobile liquid saturation on the non-Darcy flow coefficient in porous media. *SPE Production Engineering, 2*(04), 331–338. https://doi.org/10.2118/14206-PA.

Evans, R., Marconi, U. M. B., & Tarazona, P. (1986). Fluids in narrow pores - adsorption, capillary condensation, and critical-points. *Journal of Chemical Physics, 84*(4), 2376–2399. https://doi.org/10.1063/1.450352.

Fathi, E., Tinni, A., & Akkutlu, I. Y. (2012). Correction to Klinkenberg slip theory for gas flow in nano-capillaries. *International Journal of Coal Geology, 103*, 51–59. https://doi.org/10.1016/j.coal.2012.06.008.

Fenton, L. (1960). The sum of log-normal probability distributions in scatter transmission systems. *IRE Transactions on Communications Systems, 8*(1), 57–67. https://doi.org/10.1109/TCOM.1960.1097606.

Fick, A. (1855). V. On liquid diffusion. *The London, Edinburgh, and Dublin Philosophical Magazine and Journal of Science, 10*(63), 30–39. https://doi.org/10.1080/14786445508641925.

Forchheimer, P. (1901). Water movement through the ground. *Zeitschrift Des Vereines Deutscher Ingenieure, 45*, 1781–1788.

Frederick, D. C., Jr., & Graves, R. M. (1994). New correlations to predict non-Darcy flow coefficients at immobile and mobile water saturation. In *Paper presented at the SPE annual technical conference and exhibition, New Orleans, Louisiana.* https://doi.org/10.2118/28451-MS.

Freeman, C. M., Moridis, G. J., & Blasingame, T. A. (2011). A numerical study of microscale flow behavior in tight gas and shale gas reservoir systems. *Transport in Porous Media, 90*(1), 253–268. https://doi.org/10.1007/s11242-011-9761-6.

Geertsma, J. (1974). Estimating coefficient of inertial resistance in fluid-flow through porous-media. *Society of Petroleum Engineers Journal, 14*(5), 445–450. https://doi.org/10.2118/4706-Pa.

Gelb, L. D., Gubbins, K. E., Radhakrishnan, R., & Sliwinska-Bartkowiak, M. (1999). Phase separation in confined systems. *Reports on Progress in Physics, 62*(12), 1573–1659. https://doi.org/10.1088/0034-4885/62/12/201.

Ghosh, S., Rai, C. S., Sondergeld, C. H., & Larese, R. E. (2014). Experimental investigation of proppant diagenesis. In *Paper presented at the SPE/CSUR unconventional resources conference – Canada, Calgary, Alberta, Canada.* https://doi.org/10.2118/171604-MS.

Gor, G. Y., Paris, O., Prass, J., Russo, P. A., Carrott, M. M. L. R., & Neimark, A. V. (2013). Adsorption of n-pentane on mesoporous silica and adsorbent deformation. *Langmuir, 29*(27), 8601–8608. https://doi.org/10.1021/la401513n.

Green, L., & Duwez, P. (1951). Fluid flow through porous metals. *Journal of Applied Mechanics-transactions of the Asme, 18*(1), 39–45.

Grigg, R. B., & Hwang, M. K. (1998). High velocity gas flow effects in porous gas-water system. In *Paper presented at the SPE gas technology symposium, Calgary, Alberta, Canada.* https://doi.org/10.2118/39978-MS.

Gubbins, K. E., Long, Y., & Sliwinska-Bartkowiak, M. (2014). Thermodynamics of confined nano-phases. *Journal of Chemical Thermodynamics, 74*, 169–183. https://doi.org/10.1016/j.jct.2014.01.024.

Guo, W., Hu, Z., Zhang, X., Yu, R., & Wang, L. (2017). Shale gas adsorption and desorption characteristics and its effects on shale permeability. *Energy Exploration & Exploitation, 35*(4), 463–481. https://doi.org/10.1177/0144598716684306.

Hadjiconstantinou, N. G. (2006). The limits of Navier-Stokes theory and kinetic extensions for describing small-scale gaseous hydrodynamics. *Physics of Fluids, 18*(11). https://doi.org/10.1063/1.2393436.

Hamada, Y., Koga, K., & Tanaka, H. (2007). Phase equilibria and interfacial tension of fluids confined in narrow pores. *Journal of Chemical Physics, 127*(8). https://doi.org/10.1063/1.2759926.

Hassan, M. H. M., & Douglas Way, J. (1996). Gas transport in a microporous silica membrane. In *Paper presented at the Abu Dhabi international petroleum exhibition and conference, Abu Dhabi, United Arab Emirates.* https://doi.org/10.2118/36226-MS.

Heller, R., & Zoback, M. (2014). Adsorption of methane and carbon dioxide on gas shale and pure mineral samples. *Journal of Unconventional Oil and Gas Resources, 8*, 14–24. https://doi.org/10.1016/j.juogr.2014.06.001.

Hirschfelder, J. O., Curtiss, C. F., Bird, R. B., & University of Wisconsin Theoretical Chemistry Laboratory. (1954). *Molecular theory of gases and liquids.* Wiley.

Hornyak, G. L., Tibbals, H. F., Dutta, J., & Moore, J. J. (2008). *Introduction to nanoscience and nanotechnology.* CRC Press.

Hubbert, M. K. (1956). Darcys law and the field equations of the flow of underground fluids. *Transactions of the American Institute of Mining and Metallurgical Engineers, 207*(10), 223–239.

Irmay, S. (1958). On the theoretical derivation of Darcy and Forchheimer formulas. *Eos, Transactions American Geophysical Union, 39*(4), 702–707. https://doi.org/10.1029/TR039i004p00702.

Islam, A., & Patzek, T. (2014). Slip in natural gas flow through nanoporous shale reservoirs. *Journal of Unconventional Oil and Gas Resources, 7*, 49–54. https://doi.org/10.1016/j.juogr.2014.05.001.

Janicek, J. D., & Katz, D. L. (1955). *Applications of unsteady state gas flow calculations.*

Javadpour, F. (2009). Nanopores and apparent permeability of gas flow in mudrocks (shales and siltstone). *Journal of Canadian Petroleum Technology, 48*(8), 16–21. https://doi.org/10.2118/09-08-16-Da.

Javadpour, F., Fisher, D., & Unsworth, M. (2007). Nanoscale gas flow in shale gas sediments. *Journal of Canadian Petroleum Technology, 46*(10), 55–61. https://doi.org/10.2118/07-10-06.

Jin, L., Ma, Y., & Ahmad, J. (2013). Investigating the effect of pore proximity on phase behavior and fluid properties in shale formations. In *Paper presented at the SPE annual technical conference and exhibition, New Orleans, Louisiana, USA.* https://doi.org/10.2118/166192-MS.

Jones, S. C. (1987). Using the inertial coefficient, B, to characterize heterogeneity in reservoir rock. In *Paper presented at the SPE annual technical conference and exhibition, Dallas, Texas.* https://doi.org/10.2118/16949-MS.

Karniadakis, G. E., Beskok, A., & Aluru, N. (2005). *Microflows and nanoflows: Fundamentals and simulation.* New York: Springer.

de Keizer, A., Michalski, T., & Findenegg, G. H. (1991). Fluids in pores: Experimental and computer simulation studies of multilayer adsorption, pore condensation and critical-point shifts. In *Pure and applied chemistry.*

Keller, J. U., & Staudt, R. (2005). *Gas adsorption equilibria: Experimental methods and adsorptive isotherms.* Springer.

Kim, T. H. (2018). *Integrative modeling of CO_2 injection for enhancing hydrocarbon recovery and CO_2 storage in shale reservoirs.* Hanyang University.

Kim, T. H., Cho, J., & Lee, K. S. (2017). Evaluation of CO_2 injection in shale gas reservoirs with multi-component transport and geomechanical effects. *Applied Energy, 190*, 1195–1206. https://doi.org/10.1016/j.apenergy.2017.01.047.

Klinkenberg, L. J. (1941). The permeability of porous media to liquids and gases. In *Paper presented at the drilling and production practice, New York, New York.*

Krishna, R. (1993). Problems and pitfalls in the use of the Fick formulation for intraparticle diffusion. *Chemical Engineering Science, 48*(5), 845–861. https://doi.org/10.1016/0009-2509(93)80324-J.

Kruk, M., Jaroniec, M., Ko, C. H., & Ryoo, R. (2000). Characterization of the porous structure of SBA-15. *Chemistry of Materials, 12*(7), 1961–1968. https://doi.org/10.1021/cm000164e.

Kruk, M., Jaroniec, M., & Sayari, A. (1997). Adsorption study of surface and structural properties of MCM-41 materials of different pore sizes. *Journal of Physical Chemistry B, 101*(4), 583–589. https://doi.org/10.1021/jp962000k.

Kruk, M., Jaroniec, M., & Sayari, A. (1999). Relations between pore structure parameters and their implications for characterization of MCM-41 using gas adsorption and X-ray diffraction. *Chemistry of Materials, 11*(2), 492–500. https://doi.org/10.1021/cm981006e.

Kutasov, I. M. (1993). Equation predicts non-Darcy flow coefficient. *Oil & Gas Journal, 91*(11), 66–67.

Langmuir, I. (1918). The adsorption of gases on plane surfaces of glass, mica and platinum. *Journal of the American Chemical Society, 40*, 1361–1403. https://doi.org/10.1021/ja02242a004.

Li, D., & Engler, T. W. (2001). Literature review on correlations of the non-Darcy coefficient. In *Paper presented at the SPE permian basin oil and gas recovery conference, Midland, Texas.* https://doi.org/10.2118/70015-MS.

Li, B., Mehmani, A., Chen, J., Georgi, D. T., & Jin, G. (2013). The condition of capillary condensation and its effects on adsorption isotherms of unconventional gas condensate reservoirs. In *Paper presented at the SPE annual technical conference and exhibition, New Orleans, Louisiana, USA.* https://doi.org/10.2118/166162-MS.

Li, D., Svec, R. K., Engler, T. W., & Grigg, R. B. (2001). Modeling and simulation of the wafer non-Darcy flow experiments. In *Paper presented at the SPE western regional meeting, Bakersfield, California.* https://doi.org/10.2118/68822-MS.

Liu, X., Civan, F., & Evans, R. D. (1995). Correlation of the non-Darcy flow coefficient. *Journal of Canadian Petroleum Technology, 34*(10), 50–54. https://doi.org/10.2118/95-10-05.

Li, Z. Y., Zhang, X. X., & Liu, Y. (2017). Pore-scale simulation of gas diffusion in unsaturated soil aggregates: Accuracy of the dusty-gas model and the impact of saturation. *Geoderma, 303*, 196–203. https://doi.org/10.1016/j.geoderma.2017.05.008.

Long, Y., Sliwinska-Bartkowiak, M., Drozdowski, H., Kempinski, M., Phillips, K. A., Palmer, J. C., et al. (2013). High pressure effect in nanoporous carbon materials: Effects of pore geometry. *Colloids and Surfaces A-Physicochemical and Engineering Aspects, 437*, 33–41. https://doi.org/10.1016/j.colsurfa.2012.11.024.

Lu, X. C., Li, F. C., & Watson, A. T. (1995a). Adsorption measurements in devonian shales. *Fuel, 74*(4), 599–603. https://doi.org/10.1016/0016-2361(95)98364-K.

Lu, X. C., Li, F. C., & Watson, A. T. (1995b). Adsorption studies of natural-gas storage in devonian shales. *SPE Formation Evaluation, 10*(2), 109–113. https://doi.org/10.2118/26632-Pa.

Macdonald, I. F., Elsayed, M. S., Mow, K., & Dullien, F. A. L. (1979). Flow through porous-media — Ergun equation revisited. *Industrial & Engineering Chemistry Fundamentals, 18*(3), 199–208. https://doi.org/10.1021/i160071a001.

Mason, E. A., & Malinauskas, A. P. (1983). *Gas transport in porous media: The dusty-gas model.* Elsevier.

Maxwell, J. C. (1995). *The scientific letters and papers of James Clerk Maxwell.* Cambridge University Press.

Mengal, S. A., & Wattenbarger, R. A. (2011). Accounting for adsorbed gas in shale gas reservoirs. In *Paper presented at the SPE Middle East oil and gas show and conference, Manama, Bahrain.* https://doi.org/10.2118/141085-MS.

Mitropoulos, A. C. (2008). The Kelvin equation. *Journal of Colloid and Interface Science, 317*(2), 643–648. https://doi.org/10.1016/j.jcis.2007.10.001.

Moghanloo, R. G., Yuan, B., Ingrahama, N., Krampf, E., Arrowooda, J., & Dadmohammadi, Y. (2015). Applying macroscopic material balance to evaluate interplay between dynamic drainage volume and well performance in tight formations. *Journal of Natural Gas Science and Engineering, 27*, 466–478. https://doi.org/10.1016/j.jngse.2015.07.047.

Morishige, K., & Nobuoka, K. (1997). X-ray diffraction studies of freezing and melting of water confined in a mesoporous adsorbent (MCM-41). *Journal of Chemical Physics, 107*(17), 6965–6969. https://doi.org/10.1063/1.474936.

Mower, M. B. (2005). *Competitive desorption of carbon tetrachloride + water from mesoporous silica particles.* Thesis (M.S. in chemical engineering). Washington State University. August 2005.

Myong, R. S. (2001). A computational method for Eu's generalized hydrodynamic equations of rarefied and microscale gasdynamics. *Journal of Computational Physics, 168*(1), 47–72. https://doi.org/10.1006/jcph.2000.6678.

Myong, R. S. (2004). Gaseous slip models based on the Langmuir adsorption isotherm. *Physics of Fluids, 16*(1), 104–117. https://doi.org/10.1063/1.1630799.

Naraghi, M. E., & Javadpour, F. (2015). A stochastic permeability model for the shale-gas systems. *International Journal of Coal Geology, 140*, 111–124. https://doi.org/10.1016/j.coal.2015.02.004.

Naumov, S., Valiullin, R., Monson, P. A., & Karger, J. (2008). Probing memory effects in confined fluids via diffusion measurements. *Langmuir, 24*(13), 6429–6432. https://doi.org/10.1021/la801349y.

Nelson, P. H. (2009). Pore-throat sizes in sandstones, tight sandstones, and shales. *Aapg Bulletin, 93*(3), 329–340. https://doi.org/10.1306/10240808059.

Nojabaei, B., Johns, R. T., & Chu, L. (2013). Effect of capillary pressure on phase behavior in tight rocks and shales. *SPE Reservoir Evaluation & Engineering, 16*(3), 281–289. https://doi.org/10.2118/159258-Pa.

Nuttal, B. C., Eble, C., Bustin, R. M., & Drahovzal, J. A. (2005). Analysis of Devonian black shales in Kentucky for potential carbon dioxide sequestration and enhanced natural gas production. In E. S. Rubin, D. W. Keith, C. F. Gilboy, M. Wilson, T. Morris, J. Gale, et al. (Eds.), *Greenhouse gas control technologies 7* (pp. 2225–2228). Oxford: Elsevier Science Ltd.

Palmer, I., & Mansoori, J. (1998). How permeability depends on stress and pore pressure in coalbeds: A new model.

SPE Reservoir Evaluation & Engineering, 1(06), 539−544. https://doi.org/10.2118/52607-PA.

Pascal, H., & Quillian, R. G. (1980). Analysis of vertical fracture length and non-Darcy flow coefficient using variable rate tests. In *Paper presented at the SPE annual technical conference and exhibition, Dallas, Texas.* https://doi.org/10.2118/9348-MS.

Pedrosa, O. A., Jr. (1986). Pressure transient response in stress-sensitive formations. In *Paper presented at the SPE California regional meeting, Oakland, California.* https://doi.org/10.2118/15115-MS.

Radhakrishnan, R., Gubbins, K. E., & Sliwinska-Bartkowiak, M. (2002). Global phase diagrams for freezing in porous media. *Journal of Chemical Physics, 116*(3), 1147−1155. https://doi.org/10.1063/1.1426412.

Raghavan, R., & Chin, L. Y. (2004). Productivity changes in reservoirs with stress-dependent permeability. *SPE Reservoir Evaluation & Engineering, 7*(4), 308−315. https://doi.org/10.2118/88870-Pa.

Rathakrishnan, E. (2004). *Gas dynamics.* Prentice Hall of India private limited.

Ravikovitch, P. I., & Neimark, A. V. (2001). Characterization of nanoporous materials from adsorption and desorption isotherms. *Colloids and Surfaces A-Physicochemical and Engineering Aspects, 187,* 11−21. https://doi.org/10.1016/S0927-7757(01)00614-8.

Reid, R. C. (1977). The properties of gases and liquids/Robert C. Reid, John M. Prausnitz, Thomas K. Sherwood. In J. M. Prausnitz, & T. K. Sherwood (Eds.), *McGraw-Hill chemical engineering series.* New York: McGraw-Hill.

Rezaee, R. (2015). *Fundamentals of gas shale reservoirs.* Wiley.

Ross, D. J. K., & Bustin, R. M. (2007). Impact of mass balance calculations on adsorption capacities in microporous shale gas reservoirs. *Fuel, 86*(17−18), 2696−2706. https://doi.org/10.1016/j.fuel.2007.02.036.

Roy, S., Raju, R., Chuang, H. F., Cruden, B. A., & Meyyappan, M. (2003). Modeling gas flow through microchannels and nanopores. *Journal of Applied Physics, 93*(8), 4870−4879. https://doi.org/10.1063/1.1559936.

Russo, P. A., Carrott, M. M. L. R., & Carrott, P. J. M. (2012). Trends in the condensation/evaporation and adsorption enthalpies of volatile organic compounds on mesoporous silica materials. *Microporous and Mesoporous Materials, 151,* 223−230. https://doi.org/10.1016/j.micromeso.2011.10.032.

Sakhaee-Pour, A., & Bryant, S. L. (2012). Gas permeability of shale. *SPE Reservoir Evaluation & Engineering, 15*(4), 401−409. https://doi.org/10.2118/146944-Pa.

Samier, P., Onaisi, A., & Fontaine, G. (2003). Coupled analysis of geomechanics and fluid flow in reservoir simulation. In *Paper presented at the SPE reservoir simulation symposium, Houston, Texas.* https://doi.org/10.2118/79698-MS.

Scheidegger, A. E. (1958). The physics of flow through porous media. *Soil Science, 86*(6).

Sheng, J. J. (2015). Enhanced oil recovery in shale reservoirs by gas injection. *Journal of Natural Gas Science and Engineering, 22,* 252−259. https://doi.org/10.1016/j.jngse.2014.12.002.

Sheng, G., Javadpour, F., & Su, Y. (2018). Effect of microscale compressibility on apparent porosity and permeability in shale gas reservoirs. *International Journal of Heat and Mass Transfer, 120,* 56−65. https://doi.org/10.1016/j.ijheatmasstransfer.2017.12.014.

Sheng, G. L., Su, Y. L., Wang, W. D., Liu, J. H., Lu, M. J., Zhang, Q., et al. (2015). A multiple porosity media model for multi-fractured horizontal wells in shale gas reservoirs. *Journal of Natural Gas Science and Engineering, 27,* 1562−1573. https://doi.org/10.1016/j.jngse.2015.10.026.

Shi, J.-Q., & Durucan, S. (2008). Modelling of mixed-gas adsorption and diffusion in coalbed reservoirs. In *Paper presented at the SPE unconventional reservoirs conference, Keystone, Colorado, USA.* https://doi.org/10.2118/114197-MS.

Shim, W. G., Lee, J. W., & Moon, H. (2006). Heterogeneous adsorption characteristics of volatile organic compounds (VOCs) on MCM-48. *Separation Science and Technology, 41*(16), 3693−3719. https://doi.org/10.1080/01496390600956936.

Sigal, R. F. (2015). Pore-size distributions for organic-shale-reservoir rocks from nuclear-magnetic-resonance spectra combined with adsorption measurements. *SPE Journal, 20*(4), 824−830. https://doi.org/10.2118/174546-Pa.

Sigmund, P. M. (1976a). Prediction of molecular-diffusion at reservoir conditions .1. Measurement and prediction of binary dense gas-diffusion coefficients. *Journal of Canadian Petroleum Technology, 15*(2), 48−57. https://doi.org/10.2118/76-02-05.

Sigmund, P. M. (1976b). Prediction of molecular diffusion at reservoir conditions. Part II - estimating the effects of molecular diffusion and convective mixing in multicomponent systems. *Journal of Canadian Petroleum Technology, 15*(03), 11. https://doi.org/10.2118/76-03-07.

Silin, D., & Kneafsey, T. (2012). Shale gas: Nanometer-scale observations and well modelling. *Journal of Canadian Petroleum Technology, 51*(6), 464−475. https://doi.org/10.2118/149489-Pa.

Sing, K. S. W. (1982). Reporting physisorption data for gas solid systems − with special reference to the determination of surface-area and porosity. *Pure and Applied Chemistry, 54*(11), 2201−2218. https://doi.org/10.1351/pac198254112201.

Singh, H., & Javadpour, F. (2016). Langmuir slip-Langmuir sorption permeability model of shale. *Fuel, 164,* 28−37. https://doi.org/10.1016/j.fuel.2015.09.073.

Singh, H., Javadpour, F., Ettehadtavakkol, A., & Darabi, H. (2014). Nonempirical apparent permeability of shale. *SPE Reservoir Evaluation & Engineering, 17*(3), 414−424. https://doi.org/10.2118/170243-Pa.

Singh, S. K., Sinha, A., Deo, G., & Singh, J. K. (2009). Vapor-liquid phase coexistence, critical properties, and surface tension of confined alkanes. *Journal of Physical Chemistry C, 113*(17), 7170−7180. https://doi.org/10.1021/jp8073915.

Sinha, S., Braun, E. M., Determan, M. D., Passey, Q. R., Leonardi, S. A., Boros, J. A., et al. (2013). Steady-state permeability measurements on intact shale samples at reservoir conditions − effect of stress, temperature, pressure, and type of gas. In *Paper presented at the SPE Middle East oil and gas show and conference, Manama, Bahrain.* https://doi.org/10.2118/164263-MS.

Sondergeld, C. H., Ambrose, R. J., Rai, C. S., & Moncrieff, J. (2010). Micro-Structural studies of gas shales. In *Paper presented at the SPE unconventional gas conference, Pittsburgh, Pennsylvania, USA.* https://doi.org/10.2118/131771-MS.

Song, W. H., Yao, J., Li, Y., Sun, H., Zhang, L., Yang, Y. F., et al. (2016). Apparent gas permeability in an organic-rich shale reservoir. *Fuel, 181,* 973−984. https://doi.org/10.1016/j.fuel.2016.05.011.

Stiel, L. I., & George, T. (1962). Lennard-Jones force constants predicted from critical properties. *Journal of Chemical & Engineering Data, 7*(2), 234−236. https://doi.org/10.1021/je60013a023.

Sumner, M. E. (1999). *Handbook of soil science.* Taylor & Francis.

Tanchoux, N., Trens, P., Maldonado, D., Di Renzo, F., & Fajula, F. (2004). The adsorption of hexane over MCM-41 type materials. *Colloids and Surfaces A-Physicochemical and Engineering Aspects, 246*(1−3), 1−8. https://doi.org/10.1016/j.colsurfa.2004.05.033.

Teklu, T. W., Alharthy, N., Kazemi, H., Yin, X. L., Graves, R. M., & AlSumaiti, A. M. (2014). Phase behavior and minimum miscibility pressure in nanopores. *SPE Reservoir Evaluation & Engineering, 17*(3), 396−403. https://doi.org/10.2118/168865-Pa.

Terzaghi, K. (1943). *Theoretical soil mechanics.* J. Wiley and Sons, Inc.

Thauvin, F., & Mohanty, K. K. (1998). Network modeling of non-Darcy flow through porous media. *Transport in Porous Media, 31*(1), 19−37. https://doi.org/10.1023/A:1006558926606.

Thommes, M., & Cychosz, K. A. (2014). Physical adsorption characterization of nanoporous materials: Progress and challenges. *Adsorption-journal of the International Adsorption Society, 20*(2−3), 233−250. https://doi.org/10.1007/s10450-014-9606-z.

Thommes, M., & Findenegg, G. H. (1994). Pore condensation and critical-point shift of a fluid in controlled-pore glass. *Langmuir, 10*(11), 4270−4277. https://doi.org/10.1021/la00023a058.

Thompson, J. M., Mangha, V. O., & Anderson, D. M. (2011). Improved shale gas production forecasting using a simplified analytical method-A Marcellus case study. In *Paper presented at the North American unconventional gas conference and exhibition, The Woodlands, Texas, USA.* https://doi.org/10.2118/144436-MS.

Thomson, W. (1872). 4. On the equilibrium of vapour at a curved surface of liquid. *Proceedings of the Royal Society of Edinburgh, 7,* 63−68. https://doi.org/10.1017/S0370164600041729.

Tran, D., Buchanan, L., & Nghiem, L. (2010). Improved gridding technique for coupling geomechanics to reservoir flow. *SPE Journal, 15*(1), 64−75. https://doi.org/10.2118/115514-Pa.

Tran, D., Nghiem, L., & Buchanan, L. (2005a). An overview of iterative coupling between geomechanical deformation and reservoir flow. In *Paper presented at the SPE international thermal operations and heavy oil symposium, Calgary, Alberta, Canada.* https://doi.org/10.2118/97879-MS.

Tran, D., Nghiem, L., & Buchanan, L. (2005b). Improved iterative coupling of geomechanics with reservoir simulation. In *Paper presented at the SPE reservoir simulation symposium, The Woodlands, Texas.* https://doi.org/10.2118/93244-MS.

Tran, D., Nghiem, L., & Buchanan, L. (2009). Aspects of coupling between petroleum reservoir flow and geomechanics. In *Paper presented at the 43rd U.S. Rock mechanics symposium & 4th U.S. − Canada rock mechanics symposium, Asheville, North Carolina.*

Tran, D., Settari, A., & Nghiem, L. (2002). New iterative coupling between a reservoir simulator and a geomechanics module. In *Paper presented at the SPE/ISRM rock mechanics conference, Irving, Texas.* https://doi.org/10.2118/78192-MS.

Tran, D., Settari, A., & Nghiem, L. (2004). New iterative coupling between a reservoir simulator and a geomechanics module. *SPE Journal, 9*(3), 362−368. https://doi.org/10.2118/88989-Pa.

Travalloni, L., Castier, M., Tavares, F. W., & Sandler, S. I. (2010). Critical behavior of pure confined fluids from an extension of the van der Waals equation of state. *Journal of Supercritical Fluids, 55*(2), 455−461. https://doi.org/10.1016/j.supflu.2010.09.008.

Tzoulaki, D., Heinke, L., Lim, H., Li, J., Olson, D., Caro, J., et al. (2009). Assessing surface permeabilities from transient guest profiles in nanoporous host materials. *Angewandte Chemie-international Edition, 48*(19), 3525−3528. https://doi.org/10.1002/anie.200804785.

Veltzke, T., & Thoming, J. (2012). An analytically predictive model for moderately rarefied gas flow. *Journal of Fluid Mechanics, 698,* 406−422. https://doi.org/10.1017/jfm.2012.98.

Vermylen, J. P. (2011). *Geomechanical studies of the Barnett shale.* Texas, USA: Department of Geophysics, Stanford University.

Wang, X., Luo, P., Er, V., & Huang, S.-S. S. (2010). Assessment of CO_2 flooding potential for bakken formation, saskatchewan. In *Paper presented at the Canadian unconventional resources and international petroleum conference, Calgary, Alberta, Canada.* https://doi.org/10.2118/137728-MS.

Wang, J., Luo, H. S., Liu, H. Q., Cao, F., Li, Z. T., & Sepehrnoori, K. (2017). An integrative model to simulate gas transport and production coupled with gas adsorption, non-Darcy flow, surface diffusion, and stress dependence in organic-shale reservoirs. *SPE Journal, 22*(1), 244−264. https://doi.org/10.2118/174996-Pa.

Wasaki, A., & Akkutlu, I. Y. (2015). Permeability of organic-rich shale. *SPE Journal, 20*(6), 1384−1396. https://doi.org/10.2118/170830-Pa.

Webb, S. W., & Pruess, K. (2003). The use of Fick's law for modeling trace gas diffusion in porous media. *Transport in Porous Media, 51*(3), 327−341. https://doi.org/10.1023/A:1022379016613.

Whitney, D. D. (1988). *Characterization of the non-Darcy flow coefficient in propped hydraulic fractures.* Thesis (M.S.). University of Oklahoma.

Wilke, C. R., & Chang, P. (1955). Correlation of diffusion coefficients in dilute solutions. *Aiche Journal, 1*(2), 264–270. https://doi.org/10.1002/aic.690010222.

Wong, S. W. (1970). Effect of liquid saturation on turbulence factors for gas-liquid systems. *Journal of Canadian Petroleum Technology, 9*(4), 274.

Wu, K. L., Li, X. F., Wang, C. C., Chen, Z. X., & Yu, W. (2015). A model for gas transport in microfractures of shale and tight gas reservoirs. *Aiche Journal, 61*(6), 2079–2088. https://doi.org/10.1002/aic.14791.

Yao, J., Sun, H., Fan, D. Y., Huang, Z. Q., Sun, Z. X., & Zhagn, G. H. (2013). Transport mechanisms and numerical simulation of shale gas reservoirs. *Journal of China University of Petroleum, 37*(1), 8.

Ye, G. H., Zhou, X. G., Yuan, W. K., Ye, G. H., & Coppens, M. O. (2016). Probing pore blocking effects on multiphase reactions within porous catalyst particles using a discrete model. *Aiche Journal, 62*(2), 451–460. https://doi.org/10.1002/aic.15095.

Yuan, B., Wood, D. A., & Yu, W. Q. (2015). Stimulation and hydraulic fracturing technology in natural gas reservoirs: Theory and case studies (2012–2015). *Journal of Natural Gas Science and Engineering, 26*, 1414–1421. https://doi.org/10.1016/j.jngse.2015.09.001.

Yun, J. H., Duren, T., Keil, F. J., & Seaton, N. A. (2002). Adsorption of methane, ethane, and their binary mixtures on MCM-41: Experimental evaluation of methods for the prediction of adsorption equilibrium. *Langmuir, 18*(7), 2693–2701. https://doi.org/10.1021/la0155855.

Yu, W., & Sepehrnoori, K. (2014). Simulation of gas desorption and geomechanics effects for unconventional gas reservoirs. *Fuel, 116*, 455–464. https://doi.org/10.1016/j.fuel.2013.08.032.

Yu, W., Lashgari, H. R., Wu, K., & Sepehrnoori, K. (2015). CO_2 injection for enhanced oil recovery in Bakken tight oil reservoirs. *Fuel, 159*, 354–363. https://doi.org/10.1016/j.fuel.2015.06.092.

Yu, W., Sepehrnoori, K., & Patzek, T. W. (2016). Modeling gas adsorption in Marcellus shale with Langmuir and BET isotherms. *SPE Journal, 21*(2), 589–600. https://doi.org/10.2118/170801-Pa.

Yu, W., Sepehrnoori, K., & Wiktor Patzek, T. (2014). Evaluation of gas adsorption in Marcellus shale. In *Paper presented at the SPE annual technical conference and exhibition, Amsterdam, The Netherlands*. https://doi.org/10.2118/170801-MS.

Zarragoicoechea, G. J., & Kuz, V. A. (2004). Critical shift of a confined fluid in a nanopore. *Fluid Phase Equilibria, 220*(1), 7–9. https://doi.org/10.1016/j.fluid.2004.02.014.

Zeng, Y., Ning, Z., Lei, Y., Huang, L., Lv, C., & Hou, Y. (2017). Analytical model for shale gas transportation from matrix to fracture network. In *Paper presented at the SPE europec featured at 79th EAGE conference and exhibition, Paris, France*. https://doi.org/10.2118/185794-MS.

Zhang, Y., Civan, F., Devegowda, D., & Sigal, R. F. (2013). Improved prediction of multi-component hydrocarbon fluid properties in organic rich shale reservoirs. In *Paper presented at the SPE annual technical conference and exhibition, New Orleans, Louisiana, USA*. https://doi.org/10.2118/166290-MS.

Zhang, M., Yao, J., Sun, H., Zhao, J. L., Fan, D. Y., Huang, Z. Q., et al. (2015). Triple-continuum modeling of shale gas reservoirs considering the effect of kerogen. *Journal of Natural Gas Science and Engineering, 24*, 252–263. https://doi.org/10.1016/j.jngse.2015.03.032.

Zheng, D., Yuan, B., Rouzbeh, G., & Moghanloo. (2017). Analytical modeling dynamic drainage volume for transient flow towards multi-stage fractured wells in composite shale reservoirs. *Journal of Petroleum Science and Engineering, 149*, 756–764. https://doi.org/10.1016/j.petrol.2016.11.023.

CHAPTER 4

Simulation of Shale Reservoirs

ABSTRACT

Although shale industry has mostly concentrated on the hydraulic fractures for modeling of shale reservoir, various transport mechanisms should also be considered. This chapter presents comprehensive methodology and process of shale reservoir simulation. Matrix-fracture system can be constructed by dual porosity and dual permeability models. Hydraulic fractures can be generated by planar model with logarithmically spaced, locally refined grids and complex models based on microseismic data. Shale reservoir models consider non-Darcy flow with Forchheimer equation, desorption with Langmuir and BET isotherms, geomechanics with exponential and power law correlations, and phase behavior shift in nanopores with critical point shift method. Field examples of shale gas and oil reservoirs are presented with these mechanisms and overall modeling processes.

NUMERICAL MODELING OF SHALE RESERVOIRS

Modeling of Natural Fracture System

Extensive studies for a numerical model of shale reservoirs have been presented since shale boom (Anderson, Nobakht, Moghadam, & Mattar, 2010; Cipolla, Lolon, Erdle, & Rubin, 2010; Kam, Nadeem, Novlesky, Kumar, & Omatsone, 2015; Novlesky, Kumar, & Merkle, 2011; Rubin, 2010; Yu, 2015). For accurate modeling of shale reservoirs, several numerical methodologies to realize the characteristics of shale reservoirs should be fully understood. One of the essential features of shale reservoirs is the matrix-fracture system. Because of the low permeability of shale matrix, natural fracture system contributes significantly to the production in shale reservoirs. Therefore, the modeling of the matrix-fracture system is of importance to predict and improve oil and gas production. Generally, the matrix-fracture system is calculated by the dual porosity model because of the different fluid storage and conductivity characteristics of the matrix and fractures as mentioned in Chapter 2.

Barenblatt, Zheltov, and Kochina (1960) and Warren and Root (1963) introduced the concept of dual porosity medium, which considers matrix and fracture as a different porous medium system. In both papers, the transfer per unit bulk volume between the matrix and the fracture was assumed to take place under pseudosteady state conditions (Lim & Aziz, 1995). Flow equations of dual porosity approach to model naturally fractured reservoirs were presented by Kazemi, Merrill, Porterfield, and Zeman (1976). Kazemi et al. (1976) assumed that fractures are orthogonal in the three directions and act as boundaries to matrix elements. If fractures provide the main path for fluid flow from the reservoir, the fluid from the matrix blocks usually flows into the fracture space, and the fractures carry the fluid to the wellbore.

The following equations are governing flow equations for dual porosity approach to model the natural fracture reservoirs (CMG, 2017b). Flow equations of dual porosity model are an extension of a single porosity model described in Collins, Nghiem, Li, and Grabonstotter (1992). Flow equations of hydrocarbon (Eq. 4.1) and water (Eq. 4.2) components and volume consistency equation (Eq. 4.3) in matrix blocks are shown below:

$$-\tau_{iomf} - \tau_{igmf} - \frac{V}{\Delta t}\left(N_i^{n+1} - N_i^n\right)_m = 0, \ i = 1, ..., n_c, \quad (4.1)$$

$$-\tau_{wmf} - \frac{V}{\Delta t}\left(N_{n_c+1}^{n+1} - N_{n_c+1}^n\right)_m = 0, \quad (4.2)$$

$$\sum_{i=1}^{n_c+1} N_{im}^{n+1} - \phi_m^{n+1}\left(\rho_o S_o + \rho_g S_g + \rho_w S_w\right)_m^{n+1} = 0, \quad (4.3)$$

where τ_{iomf} is the matrix-fracture transfer in the oil phase for the component i, τ_{igmf} the matrix-fracture transfer in the gas phase for the component i, V the grid block volume, Δt the time step, N_i the moles of the component i per unit of grid block volume, τ_{wmf} the matrix-fracture transfer for water, N_{n_c+1} the moles of water per unit of grid block volume. The subscript i with $i = 1, ..., n_c$ corresponds to the hydrocarbon component, and the subscript $n_c + 1$ denotes the water component. The superscripts n and $n + 1$ indicate the old and current time levels, respectively. The subscripts m and f

Transport in Shale Reservoirs. https://doi.org/10.1016/B978-0-12-817860-7.00004-0

correspond to the matrix and fracture, respectively. The volume consistency equation is the sum of the phase volumes per unit reservoir volume to the porosity, and it relates the pressure and molar densities in each grid block (Collins et al., 1992). The volume consistency equation involves only variable values in the grid block in question and at the $n + 1$ time level. The following equations describe the hydrocarbon flow (Eq. 4.4) and water (Eq. 4.5) components and volume consistency (Eq. 4.6) in fracture blocks for dual porosity approach:

$$\Delta T^s_{of} \gamma^s_{iof} \left(\Delta p^{n+1} - \gamma^s_o \Delta D\right)_f + \Delta T^s_{gf} \gamma^s_{igf} \left(\Delta p^{n+1} + \Delta p^s_{cog} - \gamma^s_g \Delta D\right)_f$$

$$+ q^{n+1}_i + \tau_{iomf} + \tau_{igmf} - \frac{V}{\Delta t}\left(N^{n+1}_i - N^n_i\right)_f = 0, \quad i = 1, \dots, n_c, \tag{4.4}$$

$$\Delta T^s_{wf}\left(\Delta p^{n+1} - \Delta p^s_{cwo} - \gamma^s_w \Delta D\right)_f + q^{n+1}_w + \tau_{wmf}$$

$$- \frac{V}{\Delta t}\left(N^{n+1}_{n_c+1} - N^n_{n_c+1}\right)_f = 0, \tag{4.5}$$

$$\sum^{n_c+1}_{i=1} N^{n+1}_{if} - \phi^{n+1}_f \left(\rho_o S_o + \rho_g S_g + \rho_w S_w\right)^{n+1}_f = 0, \tag{4.6}$$

where T_j the transmissibility of phase j, γ_{ij} the mole fraction of component i in phase j, γ_j the gradient of phase j, D the depth, p_{cog} the oil-gas capillary pressure, and p_{cwo} the water-oil capillary pressure. The subscripts j indicate phase of oil, gas, and water, presented by o, g, and w. The superscript s refers to n for explicit blocks and to $n + 1$ for implicit blocks.

As mentioned earlier, in dual porosity model, which was presented by Warren and Root (1963), fractures are the only pathway connected to the wellbore. The matrix of dual porosity system is not connected to the wellbore directly, and the fluid in the matrix is transported to the well through the fractures. Dual permeability system is similar to the dual porosity system except that matrix blocks of dual permeability system have one more channel for fluids flow than those of dual porosity system. Dual permeability system assumes that both matrix and fractures are connected to the wellbore directly. The fluid could flow from the fracture and matrix to the wellbore as well as travel between the matrix and fractures at the same time. Flow equation of fractures for dual permeability model are the same with those for dual porosity model, and the following equations describe the flow of hydrocarbon (Eq. 4.7) and water (Eq. 4.8) components in matrix blocks for dual permeability system:

$$\Delta T^s_{om} \gamma^s_{iom}\left(\Delta p^{n+1} - \gamma^s_o \Delta D\right)_m + \Delta T^s_{gm} \gamma^s_{igm}\left(\Delta p^{n+1} + \Delta p^s_{cog} - \gamma^s_g \Delta D\right)_m$$

$$- \tau_{iomf} - \tau_{igmf} - \frac{V}{\Delta t}\left(N^{n+1}_i - N^n_i\right)_m = 0, \quad i = 1, \dots, n_c, \tag{4.7}$$

$$\Delta T^s_{wm}\left(\Delta p^{n+1} - \Delta p^s_{cwo} - \gamma^s_w \Delta D\right)_m - \tau_{wmf}$$

$$- \frac{V}{\Delta t}\left(N^{n+1}_{n_c+1} - N^n_{n_c+1}\right)_m = 0, \tag{4.8}$$

In the above equations, the matrix-fracture transfer can be calculated by several methods depending on considerations of physical phenomena. According to Kazemi et al. (1976), the matrix-fracture transfer is given below:

$$\tau_{jmf} = \sigma V \frac{k_j \rho_j}{\mu_j}\left(p_{jm} - p_{jf}\right), \quad j = o, g, w, \tag{4.9}$$

where σ is the transfer coefficient or shape factor. In this transfer equation, the matrix and fracture blocks are assumed to be at the same depth, and gravity term is not considered. CMG (2017b) presented matrix-fracture transfer equations including gravity effects. They assume complete gravity segregation of the oil, gas, and water phases:

$$\tau_{omf} = \sigma V \frac{k_o \rho_o}{\mu_o}\left(p_{om} - p_{of}\right), \tag{4.10}$$

$$\tau_{gmf} = \sigma V \frac{k_g \rho_g}{\mu_g}\left[\begin{array}{c} p_{gm} - p_{gf} + \left(\dfrac{S_g}{1 - S_{org} - S_{wr}} - \dfrac{1}{2}\right)_m \Delta\gamma_{ogm}h \\[2mm] - \left(\dfrac{S_g}{1 - S_{org} - S_{wr}} - \dfrac{1}{2}\right)_f \Delta\gamma_{ogf}h \end{array}\right], \tag{4.11}$$

$$\tau_{wmf} = \sigma V \frac{k_w \rho_w}{\mu_w}\left[\begin{array}{c} p_{wm} - p_{wf} + \left(\dfrac{1}{2} - \dfrac{S_w - S_{wr}}{1 - S_{orw} - S_{wr}}\right)_m \Delta\gamma_{wom}h \\[2mm] + \left(\dfrac{1}{2} - \dfrac{S_w - S_{wr}}{1 - S_{orw} - S_{wr}}\right)_f \Delta\gamma_{wof}h \end{array}\right], \tag{4.12}$$

where

$\Delta\gamma_{og}$=(oil mass density−gas mass density)g,

$\Delta\gamma_{wo}$ = (water mass density−oil mass density)g,

and h is the height of the matrix element in the direction of gravity. In Eqs. (4.10)−(4.12), the capillary pressures and gravity effects are computed separately and later summed together. Following equations, consider a more rigorous approach that uses the pseudocapillary pressures. Pseudocapillary pressures can be derived with the assumption of instantaneous vertical gravity-capillary equilibrium as follows:

$$\int^{\tilde{p}_{cog}+\gamma_{og}h/2}_{\tilde{p}_{cog}-\gamma_{og}h/2} S_g dp_{cog} - \gamma_{og}hS_g = 0, \tag{4.13}$$

$$\int^{\tilde{p}_{cwo}+\gamma_{wo}h/2}_{\tilde{p}_{cwo}-\gamma_{wo}h/2} S_w dp_{cwo} - \gamma_{wo}hS_w = 0, \tag{4.14}$$

where $\tilde{p}_{cog}$ is the pseudo oil-gas capillary pressure, and $\tilde{p}_{cwo}$ is the pseudo water-oil capillary pressure. Pseudo-capillary pressures account for both the gravity and capillary effects. With these pseudocapillary pressures, the matrix-fracture transfer can be derived as shown:

$$\tau_{omf} = \sigma V \frac{k_o \rho_o}{\mu_o} (p_{om} - p_{of}), \tag{4.15}$$

$$\tau_{gmf} = \sigma V \frac{k_g \rho_g}{\mu_g} \left(p_{gm} + \tilde{p}_{cog,m} - p_{gf} - \tilde{p}_{cog,f} \right), \tag{4.16}$$

$$\tau_{wmf} = \sigma V \frac{k_w \rho_w}{\mu_w} \left(p_{wm} + \tilde{p}_{cwo,m} - p_{wf} - \tilde{p}_{cwo,f} \right), \tag{4.17}$$

As shown in Eqs. (4.13) and (4.14), the pseudocapillary pressures are functions of fluid saturations and densities as well as capillary pressures. The pseudocapillary pressures change dynamically with the variation of fluid saturations and densities. In addition, all transfer equations mentioned earlier assumes that the transfer of all phases occurs through the entire surfaces of the matrix element. If the matrix element is partially immersed with fluids, more advanced equations should be derived. Eqs. (4.18–4.20) were obtained by the modification of Eqs. (4.15–4.17) to account for partially immersed matrix as follows:

$$\tau_{omf} = \sigma V \frac{k_o \rho_o}{\mu_o} (p_{om} - p_{of}), \tag{4.18}$$

$$\tau_{gmf} = \sigma V \frac{k_g \rho_g}{\mu_g} \left\{ (p_{gm} - p_{gf}) + \left[S_{gm} + \frac{\sigma_z}{\sigma} \left(\frac{1}{2} - S_{gm} \right) \right] \right.$$
$$\left. \times \left(\tilde{p}_{cog,m} - \tilde{p}_{cog,f} \right) \right\}, $$
$$\tag{4.19}$$

$$\tau_{wmf} = \sigma V \frac{k_w \rho_w}{\mu_w} \left\{ \begin{array}{l} (p_{wm} - p_{wf}) - (p_{cwo,m} - p_{cwo,f}) \\ -\left(\frac{1}{2} \frac{\sigma_z}{\sigma} \right) \left[\left(\tilde{p}_{cwo,m} - \tilde{p}_{cwo,f} \right) - (p_{cwo,m} - p_{cwo,f}) \right] \end{array} \right\}, $$
$$\tag{4.20}$$

In the above equations, calculation of shape factor, σ, is significant. According to Warren and Root (1963), the shape factor describes the communication between matrix and fracture regions. It reflects the geometry of the matrix element and controls the flow between two porous regions. Shape factor has the dimension of the reciprocal area. Various researchers have provided shape factor formulas. Many of them are different so that there is confusion about which method is correct and should be used. Lim and Aziz (1995) and Mora and Wattenbarger (2009) presented a summary of existing equations for shape factors. The Warren and Root approach (1963) for shape factor

assumes uniformly spaced fractures and allows variations in the fracture width to satisfy the conditions of anisotropy. Warren and Root obtained the following equation:

$$\sigma = \frac{4n(n+2)}{L^2}, \tag{4.21}$$

According to Eq. (4.21), L is the spacing between the fractures, and n is 1, 2, or 3 parallel sets of fractures and is associated with different flow geometries such as slabs, columns, and cubes, respectively. Substituting values for n, and assuming equal spacing between fractures, $L_x = L_y = L_z = L$, σ is equal to $\frac{12}{L^2}$, $\frac{32}{L^2}$, and $\frac{60}{L^2}$ for 1, 2, and 3 sets of normal parallel fractures, respectively. The most widely used formula for σ was presented by Kazemi et al. (1976). It was supposedly developed by finite difference methods for a three-dimensional numerical simulator for fractured reservoirs.

$$\sigma = 4 \left(\frac{1}{L_x^2} + \frac{1}{L_y^2} + \frac{1}{L_z^2} \right), \tag{4.22}$$

In the shale reservoir models presented in following sections, Eq. (4.22) is used for dual porosity and dual permeability models. According to this equation, for equal fracture spacing, σ has a value of $\frac{4}{L^2}$, $\frac{8}{L^2}$, and $\frac{12}{L^2}$ for 1, 2, and 3 sets of fractures, respectively.

Coats (1989) derived values for σ under pseudosteady state conditions. These values are equal to $\frac{12}{L^2}$, $\frac{28.45}{L^2}$, and $\frac{49.58}{L^2}$ for 1, 2, and 3 sets of normal parallel fractures, respectively. Zimmerman, Chen, Hadgu, and Bodvarsson (1993) presented a different approach for σ values, using different flow geometries with constant pressure boundary conditions. Lim and Aziz (1995) presented analytic solutions of pressure diffusion for draining into a constant fracture pressure. A general Eq. (4.23) for shape factor was derived from Lim and Aziz's study:

$$\sigma = \pi^2 \left(\frac{1}{L_x^2} + \frac{1}{L_y^2} + \frac{1}{L_z^2} \right), \tag{4.23}$$

For equal fracture spacing, σ is equal to $\frac{3\pi^2}{L^2}$, $\frac{\pi^2}{L^2}$, and $\frac{2\pi^2}{L^2}$ for 1, 2, and 3 sets of fractures, respectively. When results from different sources are compared, real differences for σ values are noticed. Table 4.1 summarizes the shape factor values for different flow geometries according to the various authors when fracture spacing is same (Mora & Wattenbarger, 2009).

Modeling of Hydraulic Fractures

In a practical field situation, the most critical factor for the success of shale reservoir development is hydraulic

TABLE 4.1
Shape Factor Values for Different Flow
Geometries According to the Various Authors

Author	Slap	Columns	Cube
Warren and Root (1963)	$\frac{12}{L^2}$	$\frac{32}{L^2}$	$\frac{60}{L^2}$
Kazemi et al. (1976)	$\frac{4}{L^2}$	$\frac{8}{L^2}$	$\frac{12}{L^2}$
Lim and Aziz (1995)	$\frac{\pi^2}{L^2}$	$\frac{2\pi^2}{L^2}$	$\frac{3\pi^2}{L^2}$

fracturing. Beginning of shale boom came from the successful application of hydraulic fracturing technology. Hydraulic fracturing is the process of pumping fluid into a wellbore at a high injection rate to break the tight formation. The resistance to flow in the formation increases during injection, the pressure in the wellbore increases to a value called the breakdown pressure, that is the sum of the in situ compressive stress and the strength of the formation. Once the formation is broken down, a fracture is formed, and the injected fluid flows through it. From limited active perforations, the vertical fracture is created and propagates into the reservoir with two wings being 180 degrees apart. Depending on reservoir characteristics, especially in naturally fractured or cleated formations, it is possible that multiple fractures are created, and the two wings evolve in a tree-like pattern with increasing number of branches away from the injection point. After pumping fluid is injected until the fracture is wide enough to achieve target length, the fluid including proppant is injected. The purpose of the proppant is to keep apart the fracture surfaces once the pumping operation ceases. After pumping stops, because the pressure in the fracture decreases below the compressive in situ stress trying to close the fracture, proppant should be injected. Depending on the conditions of the reservoir, numerous types of proppants are used to hold open the fracture. Productivity and economic feasibility of shale reservoirs development are in reliance on success or failure of hydraulic fracturing. Therefore, precise simulation of hydraulic fractures is significant, and it has attracted enormous attention in both industrial and academic circles for several years.

Producing or injecting performance of hydraulically fractured well is important not only in fracture design, evaluation, and single well analysis but also in full-field reservoir modeling. Several analytical and numerical methods are available to study the effect of fracture on well performance (Aybar, 2014; Ji, Settari, Orr, & Sullivan, 2004). The classical analytic models for

hydraulically fractured reservoirs were developed to show the effects of hydraulic fracturing treatment on production. They were also capable of making practical sensitivity analysis for the reservoir and hydraulic fracture parameters. Productivity index comparison is the basis of these analytic models. McGuire and Sikora (1960) presented a fundamental type curve based on the comparison of the productivity index of a reservoir. This comparison is made according to the change of productivity index right before and after the hydraulic fracturing treatment. One can simply input the hydraulic fracture, wellbore, and reservoir parameters into his or her type curve to see the productivity index changing after the hydraulic fracturing treatment. Additionally, general conclusions can be driven by their type curve analysis, such as optimum fracture conductivity, and the theoretical maximum productivity index increment for a specific case. Prats (1961) developed a simple equation to calculate the steady state productivity index of a fractured well in a cylindrical drainage area, based on assumptions such as incompressible fluid flow, infinite fracture conductivity, and propped fracture height equal to formation height. Prats (1961) treated hydraulic fracture as enlarging well radius, the effective well radius equals to half of the half-length of the infinite-conductivity fracture. Prats (1961) further assumed that there is no pressure drop in the hydraulic fracture. The main contribution of this work is that the fracture length can be calculated from the production decline rate. Tinsley, Williams, Tiner, and Malone (1969) discussed the impact of the hydraulic fracture height on the production rate. In contrast to the other models, which assumed the hydraulic fracture and the formation height are equal, this model is applicable when the formation and hydraulic fracture height are not equal. Gringarten, Henry, Ramey, and Raghavan (1974) developed an analytical method to analyze the effect of fracture on transient pressure distribution in a well test. They documented the unsteady-state pressure distribution of a well with an infinite conductivity fracture. They reported three main flow regimes: early-time linear flow regime with the one-half slope in log-log plots, a pseudoradial flow regime, which gives a semilog straight line response, and pseudosteady state flow regime with a unit slope line. Gringarten, Ramey, and Raghavan (1975) presented the fractured well behavior-type curves with the dimensionless variables. They also proposed an equation for the fracture length calculation based on dimensionless wellbore pressure and formation permeability. Cinco, Heber. Samaniego, and Dominguez (1978) studied the transient behavior of wells with single finite conductivity fractures. They

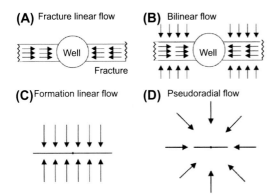

(A) Fracture linear flow

Well

Fracture

(B) Bilinear flow

Well

(C) Formation linear flow

(D) Pseudoradial flow

FIG. 4.1 Different flow regimes in hydraulically fractured reservoirs. (Credit: Nashawi, I.S., & Malallah, A.H. 2007. Well test analysis of finite-conductivity fractured wells producing at constant bottomhole pressure. *Journal of Petroleum Science and Engineering*, 57(3):303–320. https://doi.org/10.1016/j.petrol.2006.10.009.)

presented type curves, which are valid after a specific dimensionless time, and showed the differences between the finite fracture conductivity model and infinite fracture conductivity model of Gringarten et al. (1974). Cinco-Ley et al. (1978) identified four different flow regimes: linear, bilinear, formation linear, and pseudoradial flow (Fig. 4.1). Cinco-Ley and Samaniego-V (1981) proposed a new method for pressure transient analysis for fractured wells with bilinear flow regime (Fig. 4.1). Their new model detects the time at which wellbore storage effects disappear. Considering the finite fracture conductivity model, they also proposed a correlation to calculate the effective wellbore radius. This effective wellbore radius calculation is the basis of their new type curves. In addition, the reservoir parameters can be estimated by using the new type curves that are provided for bilinear flow regime. Tannich and Nierode (1985) developed a method similar to the McGuire-Sikora charts (1960) for gas wells. Bennett, Rodolfo, Reynolds, and Raghavan (1985) and Camacho-V, Raghavan, and Reynolds. (1987) presented their models for the hydraulically fractured wells producing from different layers. Bennet et al. (1985) study considered equal fracture lengths, whereas the Camacho et al. (1987) model is applicable for unequal hydraulic fracture lengths. They also stated that if two different layers are connected, it will positively affect the cumulative production performance. Cinco-Ley and Meng (1988) outlined their work on transient flow period of a well with a finite conductivity fracture. In contrast with other models, they proposed a naturally fractured reservoir with a hydraulically fractured well. To represent a dual

porosity reservoir, Cinco-Ley and Meng (1988) used pseudosteady state (Warren & Root, 1963) and transient (de Swaan, 1976; Kazemi, 1969) dual porosity models. Cinco-Ley and Meng (1988) proposed a trilinear model, which incorporates the flow in matrix blocks, natural fractures, and hydraulic fractures. They also proposed the flow regime identification method and provided the appropriate type curves for each flow regime. Bale, Smith, and Settari (1994) presented a simple, practical approach to calculating postfrac productivity for indirect vertical fracture completion wells for complex reservoir/fracture geometry.

In a complex geometry and full-field reservoir simulation, understanding the effect of hydraulic fracture on well performance is of importance. In a full-field reservoir simulation, a well is usually represented by a source/sink, and Peaceman's equations (Peaceman, 1978, 1983) are widely used to calculate the well index, which represents the strength of the source/sink. For a fractured well in reservoir simulation, the effect of fracture was simulated by using an increased well index or effective well radius, or negative skin factor (Cinco-Ley & Samaniego-V, 1981; Prats, 1961). This method is simple, but it is applicable only when the fracture is completely confined within the well block. Nghiem (1983) and Geshelin, Grabowski, and Pease (1981) modeled the infinite-conductivity vertical fracture when the fracture covers more than one block by introducing source and sink terms in the blocks communicating with a fracture. In these models, the fracture is treated as a singularity. It is assumed that elliptical flow applies in the neighborhood of the fracture and the flow into or out of the fracture is calculated from the fracture pressure and the pressures of the blocks surrounding those blocks containing the fracture. In a numerical reservoir simulation, the most rigorous method to model fracture extending over several blocks is to represent the fracture in its actual size by a plane of grid blocks. This method was first used in single-well models (Dowdle and Hyde, 1977; Holditch, 1979). However, the degree of grid refinement required is so large that it poses severe stability limitations even for fully implicit models, and the problem of grid size is even more serious in a full field model (Settari, Puchyr, & Bachman, 1990). Settari et al. (1990), Settari, Bachman, Hovem, and Paulsen (1996) represented a fracture in modeling productivity of fractured wells in single-phase models by introducing increased transmissibility for the blocks containing fracture. Al-Kobaisi, Ozkan, and Kazemi (2006) studied a hybrid reservoir model, which combines a numerical hydraulic fracture model with an analytical reservoir model. Their objective was to remove simplifying assumptions for hydraulic fracture flow. The advantage of the numerical hydraulic fracture model is that the hydraulic fracture properties such as shape, width, and variable conductivity can be modeled numerically. The reservoir flow model is still analytical and keeps the computational work manageable. Medeiros, Ozkan, and Kazemi (2006) developed a semianalytical reservoir model for horizontal wells in layered reservoirs. Reservoir heterogeneity is considered, along with local gridding, and grid boundaries are coupled analytically. Their semianalytical model is more advantageous than the analytical models when reservoir heterogeneity is considered. Brown, Ozkan, Raghavan, and Kazemi (2011) and Ozkan, Brown, Raghavan, and Kazemi (2011) presented their trilinear model. This model incorporates the flow into three different flow regions: the zone beyond the fracture tips, the area between two adjacent hydraulic fractures, and the hydraulic fracture zone. They used Laplace transformation to obtain a solution for their model. It is a common knowledge that hydraulic fracturing operation creates a stress-induced natural fracture network in tight formations, such as shale. This naturally fractured zone can be implemented in the trilinear model by using pseudosteady state or transient dual porosity models. Patzek, Frank, and Marder. (2013) presented a simplified solution for production from unconventional reservoirs. In their study, the gas diffusivity equation is solved analytically, and the pseudopressure approach is used to linearize the gas diffusivity equation.

In recent years, extensive studies for numerical modeling of hydraulic fractures have been performed to predict and evaluate production from shale reservoirs accurately (Cipolla, 2009; Cipolla, Fitzpatrick, Williams, & Ganguly, 2011; Mayerhofer, Lolon, Youngblood, & Heinze, 2006; Yu, 2015). Xu et al. (2010) developed a wire-mesh model to simulate the elliptical fracture network. However, it is difficult to simulate a nonorthogonal fracture network in the wire-mesh model. Weng, Kresse, Cohen, Wu, and Gu (2011) developed an unconventional fracture model to predict the complex fracture geometry in the formation with preexisting natural fractures using the automatic generation of unstructured grids to accurately simulate production from the complex fracture geometry (Cipolla et al., 2011; Mirzaei & Cipolla, 2012). However, this approach has some challenging practical issues such as the difficulties of model setup and long turnaround time (Zhou, Banerjee, Poe, Spath, & Thambynayagam, 2013). To investigate the effect of irregular fracture geometry on well performance of unconventional gas

reservoirs, Olorode, Freeman, George, and Blasingame (2013) proposed a 3D Voronoi mesh-generation application to generate the nonideal fracture geometry. Moinfar, Varavei, Sepehrnoori, and Johns (2013) developed an embedded discrete fracture model based on the algorithm presented by Li and Lee (2008) to simulate fluid flow from unstructured fracture geometry.

Microseismic monitoring of hydraulic fracture treatments plays a significant role in understanding the stimulation effectiveness and fracture geometry (Cipolla, 2009; Cipolla et al., 2011, Cipolla, Maxwell, Mack, 2012; Mayerhofer et al., 2006). Microseismic measurements indicate that the stimulation treatments often create complex fracture geometry, especially in the brittle shale reservoirs (Cipolla, 2009; Cipolla et al., 2011; Cipolla & Wallace, 2014; Fisher et al., 2002; Maxwell, Urbancic, Steinsberger, & Zinno, 2002; Mayerhofer et al., 2006; Warpinski, Kramm, Heinze, & Waltman, 2005). Fig. 4.2 shows the plan view of microseismic data points indicating the complex fracture geometry created in a horizontal well. The complex fracture geometry is strongly affected by insitu stresses and preexisting natural fractures (Weng, 2015; Zhou et al., 2013). Many researchers presented integrating microseismic fracture mapping with numerical modeling of hydraulic fracture networks (Cipolla, 2009; Cipolla et al., 2011; Mayerhofer et al., 2006; Novlesky et al., 2011). Although many attempts have been focused on developing hydraulic fracture models to predict the complex nonplanar fracture geometry (Weng et al. 2011; Wu, 2013, 2014; Wu, Kresse, Weng, Cohen, & Gu, 2012; Wu & Olson, 2013; Xu & Wong, 2013), it is still challenging to measure the complex fracture geometry completely and exactly because of a very complicated gridding issue, an expensive computational cost, complexities in development of computational codes, and the cost of microseismic data measurement. To overcome these challenges, Zhou et al. (2013) proposed a semianalytical model to handle the complex fracture geometry efficiently. However, the semianalytical model did not consider the effects of gas slippage, gas diffusion, gas desorption, stress-dependent fracture conductivity, and nonplanar fractures.

In most of shale reservoir fields, for the sake of simplicity, two ideal fracture geometries such as planar fractures and complex orthogonal fracture networks are widely considered to represent the geometry of hydraulic fractures for simulation of well performance (Aybar, Yu, Eshkalak, Sepehrnoori, & Patzek, 2015; Tavassoli, Yu, Javadpour, & Sepehrnoori, 2013; Yu & Sepehrnoori, 2013), as shown in Figs. 4.3 and 4.4. Planar fractures or biwing fractures intersect the well blocks containing

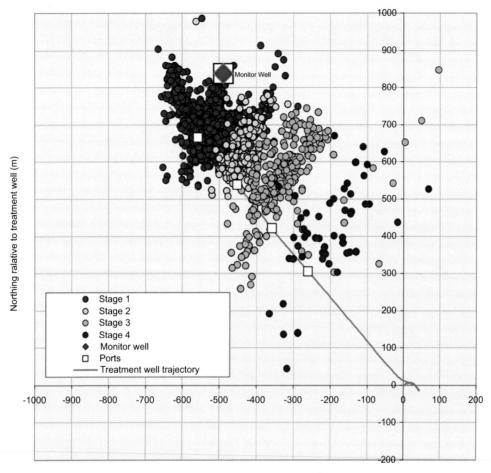

FIG. 4.2 Plan view of microseismic data points. (Credit: Novlesky, A., Kumar, A., & Merkle S. (2011). Shale gas modeling workflow: From microseismic to simulation — a Horn river case study. *Paper presented at the Canadian unconventional resources conference, Calgary, Alberta, Canada*. https://doi.org/10.2118/148710-MS.)

FIG. 4.3 Planar fracture model.

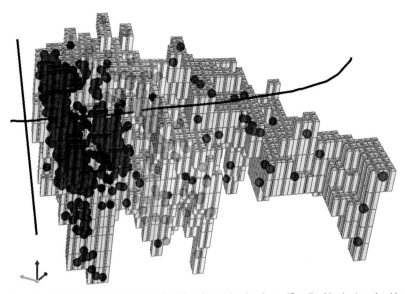

FIG. 4.4 Complex hydraulic fracture model with microseismic data. (Credit: Novlesky, A., Kumar, A., & Merkle, S. (2011). Shale gas modeling workflow: From microseismic to simulation – a Horn river case study. *Paper presented at the Canadian unconventional resources conference, Calgary, Alberta, Canada.* https://doi. org/10.2118/148710-MS.)

perforations at their midpoint. Generally, planar fractures are orthogonal to the horizontal well, and they are evenly distributed along the well. As mentioned earlier, in the past, fractures were modeled in exact width (on the order of 0.001 ft) by a plane of grid blocks. However, they use highly detailed grids and require significant time cost so that severe stability limitations are posed especially in full-field shale reservoir simulation. Rubin (2010) produced predictive fractured shale gas simulation models, which are easy to set up and that run in minutes. Based on extensive simulation works, novel models that simulate a flow inside of the stimulated reservoir volume (SRV) using coarse, logarithmically spaced, locally refined, dual permeability grids. Local grid refinement (LGR) technique is used by using numerical solutions to model hydraulic fracture explicitly with a small fracture width, which can effectively capture the transient flow behavior near the fractures (Rubin, 2010; Yu, Gao, & Sepehrnoori, 2014; Yu, Varavei, & Sepehrnoori, 2015; Yu & Sepehrnoori, 2014). Logarithmic grid spacing represents the large pressure drops near the matrix-fracture interface and reduces computation requirements for blocks further from the interface. The predictive models presented by Rubin (2010) would be small enough to be quickly run while simulating the flow in shale gas reservoir accurately. This technique allows the use of 2 ft wide

fracture conduits to mimic non-Darcy flow in 0.001 ft wide fractures.

According to several kinds of literature, in shale reservoirs, where complex fracture network is created by hydraulic fracturing, the concepts of planar fractures are insufficient to describe stimulation performance (Cipolla et al., 2010; Mayerhofer, Lolon, Warpinski, Cipolla, Walser, & Rightmire, 2010; Novlesky et al., 2011). Thus, the concept of SRV, which is the size of the created fracture network, was introduced, and microseismic data were used to calculate and model the SRV. Mayerhofer et al. (2006) and Cipolla (2009) discuss the numerical modeling of explicit fracture networks created in an SRV to simulate the physics of flow within a fractured shale reservoir. Because of the high cost of microseismic data measurement and computational cost, the application of complex fracture network models is still hard in the shale field. In recent years, planar fracture models present admittible results with the history-matching process in various shale reservoirs. Therefore, in the following section, planar fracture models were used to simulate the hydraulic fractures of shale reservoirs.

The dynamic fracture propagation models also have been presented for decades (Barree, 1983; Fanchi, Arnold, Mitchell Robert, Holstein, & Warner, 2007; Gidley and Society of Petroleum Engineers, 1989;

Green, David Barree, & Miskimins, 2007; Yousefzadeh, Qi, Virues, & Aguilera, 2017). Howard and Fast (1957) introduced the first mathematical model to design a fracture treatment. They assumed the 2D model with the constant fracture width everywhere, allowing the engineer to compute fracture area on the basis of fracture fluid leakoff characteristics of the formation and the fracturing fluid. In the following years, two of the most common 2D models were presented as simple and applicable solutions to the industry requirement for fracture designs. These 2D models are Perkins-Kern-Nordgren (PKN) and Khristianovic-Geertsma-de Klerk (KGD) models (Advani, Khattab, & Lee, 1985; Daneshy, 1973; Geertsma & De Klerk, 1969; Nolte, 1986; Nordgren, 1972; Perkins & Kern, 1961; Zheltov, 1955).

In 2D fracture modeling, the height of the fracture is considered to be constant, and width and length of fracture are calculated as a function of fracture height,

treatment parameters, and reservoir mechanical properties. Vertical propagation of fracture is confined by a change in material property of the reservoir layers or change in the horizontal minimum in situ stresses. In PKN and KGD models, the fracture deformation is considered by a linear plastic process, and it is expected that the fracture boundaries are determined in the plane of propagation. Other main prerequisites are that the injected fracturing fluid is a Newtonian fluid, the injection rate is constant, and the fracture height is fixed.

In the PKN model, the fracture planes are perpendicular to the vertical plane strain. The geometry of the PKN model is depicted in Fig. 4.5 (Adachi, Siebrits, Peirce, & Desroches, 2007). In this model, the cross-section of fracture has an elliptical shape, and it is assumed that the fracture geometry is independent of the fracture toughness (Yew & Weng, 2014). The PKN model is most appropriate when the fracture height is higher than the fracture length. Fracture length by the

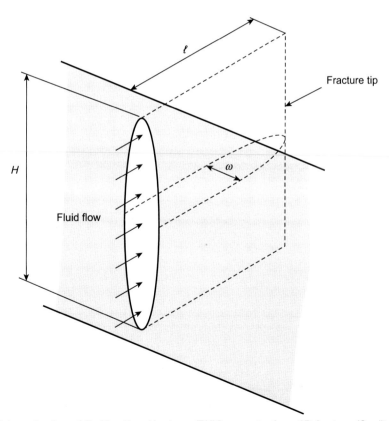

FIG. 4.5 Schematic view of Perkins-Kern-Nordgren (PKN) geometry for a 2D fracture. (Credit: Adachi, J., Siebrits, E., Peirce, A., & Desroches, J. (2007). Computer simulation of hydraulic fractures. *International Journal of Rock Mechanics and Mining Sciences, 44*(5):739—757. https://doi.org/10.1016/j.ijrmms.2006.11.006.)

PKN model without consideration of leakoff is determined by (Gidley and Society of Petroleum Engineers, 1989):

$$L_{PKN} = C_1 \left[\frac{Gq_0^3}{(1-\nu)\mu h_f^4} \right]^{\frac{1}{5}} t^{\frac{4}{5}}, \qquad (4.24)$$

where L_{PKN} is the fracture length, $C_1 = 0.45$ for two-wings fracture, G the formation shear modulus in KPa, q_0 the constant flow rate, ν the Poisson's ratio, μ the injection fluid viscosity, h_f the fracture height, and t the fracturing time.

KGD model assumes that strain is on the horizontal plane. Fig. 4.6 shows the geometry of the KGD fracture model (Adachi et al., 2007). The additional assumption for this model is that the fracture tip is cup shaped. In this model, the fracture width is constant in the vertical direction. For the KGD model in case of no leakoff to the formation and tiny dry zone at the tip of fracture, after analytic solution for fracture length calculation is achieved (Gidley and Society of Petroleum Engineers, 1989):

$$L_{KGD} = C_2 \left[\frac{Gq_0^3}{(1-\nu)\mu h_f^4} \right]^{\frac{1}{6}} t^{\frac{2}{3}}, \qquad (4.25)$$

where L_{KGD} is the fracture length and $C_2 = 0.48$ for two-wings fracture. The only difference between PKN and KGD models is the orientation of the assumed ellipse.

2D fracture propagation models were seldom found to be representative owing to the unrealistic height restrictions. With the development of high-powered computers available to most engineers, pseudo–three-dimensional (P3D) models are used by fracture design engineers (Gidley and Society of Petroleum Engineers, 1989). P3D models, which allow the fracture height to change, are the extension of 2D models. P3D models are better than 2D models for most situations because the P3D model computes the fracture height, width, and length distribution with the data for the pay zone and all the rock layers above and below the perforated interval. Gidley and Society of Petroleum Engineers, (1989) provide a detailed explanation of how P3D fracture propagation theory is used. Fig. 4.7 illustrates typical results from a P3D model. P3D models give more realistic estimates of fracture geometry and dimensions, which can lead to better designs and better description of wells. Although the complexity of P3D models increases, they are still limited by the fundamental assumptions of linear elastic deformation, the elliptical fracture shape, stress intensity factor (and singularity) at the fracture tips, and the assumptions of complete elastic coupling. These models also fail if the fracture cannot be described as a continuous entity from upper to lower tip in any element.

To model the complex geologic system of hydraulic fractures, fully 3D models have been studied as shown in Fig. 4.8 (Adachi et al., 2007; Green et al., 2007). The

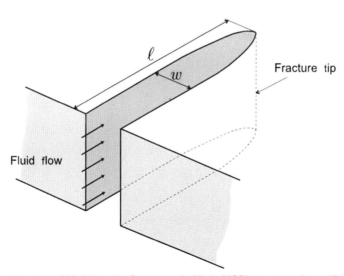

FIG. 4.6 Schematic view of Khristianovic-Geertsma-de Klerk (KGD) geometry for a 2D fracture. (Credit: Adachi, J., Siebrits, E., Peirce, A., & Desroches, J. (2007). Computer simulation of hydraulic fractures. *International Journal of Rock Mechanics and Mining Sciences, 44*(5):739–757. https://doi.org/10.1016/j. ijrmms.2006.11.006.)

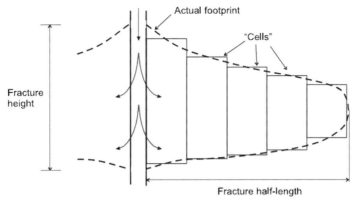

FIG. 4.7 Schematic view of fracture geometry for a P3D model. (Credit: Adachi, J., Siebrits, E., Peirce, A., & Desroches, J. (2007). Computer simulation of hydraulic fractures. *International Journal of Rock Mechanics and Mining Sciences, 44*(5):739–757. https://doi.org/10.1016/j.ijrmms.2006.11.006.)

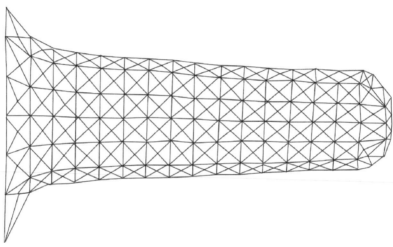

FIG. 4.8 Schematic view of fracture geometry for a 3D model based on moving mesh system of triangular elements. (Credit: Adachi, J., Siebrits, E., Peirce, A., & Desroches, J. (2007). Computer simulation of hydraulic fractures. *International Journal of Rock Mechanics and Mining Sciences, 44*(5):739–757. https://doi. org/10.1016/j.ijrmms.2006.11.006.)

main difference between the 2D/P3D and 3D models is that calculation run times are much more significant in 3D models, because of the complex calculation of all growth patterns without simplifying assumptions. Recently, several 3D fracturing simulators have been used in shale reservoirs. Barree (1983) developed a numerical simulator being capable of predicting fracture geometries during propagation in both height and length directions. The simulator is capable of handling random spatial variations in elastic properties, confining stress, pore pressure, and rock strength. Fracture fluid pressures, fracture widths, and net stresses

are calculated at uniformly spaced points over the entire fracture face. The solution of the problem combines a finite difference formulation for the calculation of fluid flow within the crack with an integral equation for fracture width. Based on a study of Barree (1983), multidisciplinary, integrated geomechanic fracture simulator GOHFER (grid-oriented hydraulic fracture extension replicator) was developed (Barree & Associates, 2015). To define fracturing design, the identification of dominant physical processes that control the fracturing process is of importance. Depending on Barree & Associates (2015), four main categories of

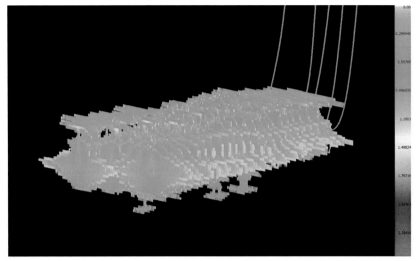

FIG. 4.9 Results of fracturing simulation in shale oil reservoir with five horizontal wells.

fracture design processes are fracture geometry creation, fluid leakoff, fluid rheology, and proppant transport. If all of these processes are considered accurately, shape, size, and conductivity of hydraulic fracture can be predicted as shown in Fig. 4.9. However, in spite of the accuracy of 3D models, various additional reservoir input data are required, and these are costly to obtain. Therefore, in most cases, to simulate the hydraulic fractures, static fracture models are used and matched with production data in shale reservoir to date. Nevertheless, the static fracture model, which is not based on complete input data and correct assumptions, cannot be a predictive model of the fracturing process. To construct a predictive model in shale reservoirs, fracture treatment should be modeled exactly with sound physics and real data.

The Process of Shale Reservoir Simulation

There have been a significant number of attempts to model the shale reservoirs (Anderson et al., 2010; Cipolla et al., 2010; Kam et al., 2015; Novlesky et al., 2011; Rubin, 2010; Yu, 2015). Although these researches on shale reservoirs have received significant interest recently, they have not reflected the combined effects of characteristic mechanisms in shale reservoirs such as non-Darcy flow, adsorption/desorption, stress-dependent compaction, and phase behavior change in nanopores. In addition, the development processes of shale resources show high cost and significant uncertainties due to many uncertain and inestimable parameters such as reservoir permeability, porosity, fracture height, fracture spacing, fracture half-length, fracture

conductivity, well spacing, and so forth. Therefore, numerical modeling with various specific mechanisms and integrative simulation approaches for sensitivity analysis, history matching, optimization, and uncertainty assessment for the economic development of shale reservoirs are valuable.

In general shale reservoir models, a dual permeability model is used to simulate the natural fracture system (Ahmed & Meehan, 2016; Novlesky et al., 2011; Rubin, 2010). The dual permeability model uses a grid block that consists of a cell to represent the matrix within a cell to express the fracture in each matrix-fracture network block so that the matrix and fracture are modeled separately. In the dual permeability model, fluid flow can occur from fracture to fracture, the matrix to fracture, and matrix to matrix grid blocks. Hydraulic fractures are modeled by the logarithmically spaced, locally refined, dual permeability methodology (Rubin, 2010). With this methodology, hydraulic fractures are explicitly modeled in the matrix portion of the dual permeability model by defining high permeability values for the hydraulic fractures and low permeability values for the shale matrix. Among refined blocks, the center blocks represent hydraulic fractures. They are set to a larger width than the real fracture width for efficient computer calculation. To conserve the hydraulic fracture conductivity in wide grid block, an effective permeability is computed as shown:

$$k_{eff} = \frac{k_f w_f}{w_{grid}}, \qquad (4.26)$$

where k_{eff} is effective fracture permeability, k_f the intrinsic fracture permeability, w_f the intrinsic fracture width, and w_{grid} the effective fracture width. Non-Darcy flow due to high velocity in the hydraulic fracture is calculated by using Forchheimer equation (1901). As mentioned in Chapter 3, Forchheimer coefficient can be calculated with various correlations. Among them, equation presented by Evans and Civan (1994) is recommended for hydraulic fracture modeling. They offered a general correlation using a large variety of data from consolidated and unconsolidated media shown next:

$$\beta = \frac{1.485 \times 10^9}{\phi k^{1.021}}. \tag{4.27}$$

Because of the increased block width, the velocity in the block will be lower than the actual velocity in the fractures so that non-Darcy flow will be miscalculated. Therefore, a non-Darcy coefficient had to be modified like fracture permeability. To compute non-Darcy effects correctly, after non-Darcy correction factor should be applied:

$$\beta_{corr} = \left(\frac{k_f}{k_{eff}}\right)^{2-N1_g} = \left(\frac{w_{grid}}{w_f}\right)^{2-N1_g}, \tag{4.28}$$

where β_{corr} is the non-Darcy correction factor and $N1_g$ is the correlation parameter. In these days, because of the high cost of microseismic measurement, planar fracture model is preferred than complex fracture model. In the models with planar fractures, hydraulic fracture spacing is significant parameters for accurate history matching.

A base numerical shale model is created using estimated reservoir properties. Because of difficulties in the measurements of reservoir and fracture parameters, many of them should be assigned reasonable ranges and matched with production data (Ahmed & Meehan, 2016; Novlesky et al., 2011). Generally, initial estimates of matrix permeability are in the range of hundreds nano-Darcy, and estimates of porosity are less than 10%. Permeability in the natural fracture system is also difficult to quantify and depends on whether the natural fractures are open, partially open, or completely healed. Initially, natural fracture permeability is assumed to be a nano-Darcy.

The properties of hydraulic fracture present the most tremendous influence on productivity in low permeability shale reservoirs. However, characteristics of hydraulic fracture systems are difficult to measure, and they cost a tremendous amount. Hydraulic fracture spacing is believed to be affected by the natural fracture spacing in the reservoir. Fracture characterization from formation image log and outcrop data with mechanical stratigraphy interpretation provides a basis for estimating a reasonable range of average fracture spacing. Hydraulic fracture spacing can be estimated from hydraulically significant natural fractures, which have large apertures enough to accept proppant. The conductivity of hydraulic fractures can be estimated by elaborate laboratory experiments (Kam et al., 2015). Shale core plugs are cut perpendicular to the bedding plane, and proppant is placed according to design between the two halves of the core to simulate an induced hydraulic fracture. The conductivity of synthetically induced fracture is measured while flowing slickwater fracturing fluid at reservoir temperature and reservoir pressure conditions. In addition, rate-transient analysis (RTA) can be used to calculate hydraulic fracture properties (Kam et al., 2015). The RTA description is used for initial assumptions when defining the induced fracture properties in the simulation model, such as height, half-length, and conductivity. RTA is representative of the behavior of the first drilled well and is valid only up to the date when the adjacent wells are drilled. It is assumed that the interaction among the adjacent wells disrupts the flow behavior, and, therefore, the RTA model is unable to account for that. Fracture design simulators, which compute geomechanic fracture propagation can also be used to estimates of hydraulic-fracture properties.

To develop a model that appropriately simulates the complex process of gas flow in shale reservoir, Kim (2018) suggested and applied several specific mechanisms such as mono- and multilayer adsorption/desorption, geomechanic deformation, confinement effects. From adsorption experiments on shale core samples, the adsorptive potential of a material depending on various pressures can be quantified at constant temperature (Heller & Zoback, 2014; Lu, Li, & Watson, 1995; Nuttal, Eble, Bustin, & Drahovzal, 2005; Ross & Bustin, 2007, 2009; Vermylen, 2011). As mentioned in Chapter 3, Langmuir isotherm (1918), which assumes single-layer adsorption covering the solid surface, is commonly used to depict adsorption/desorption behavior of methane in shale reservoirs. However, depending on Yu, Sepehrnoori, and Patzek (2016), measurements of methane adsorption deviate from the Langmuir isotherm. In Marcellus Shale core samples, data correspond with Langmuir isotherms at low pressures while data deviate from Langmuir isotherms at high pressure and fit well with BET isotherms (1938). Therefore, Eqs. (3.31) and (3.35) are used to model the Langmuir and BET adsorption as shown:

$$V = \frac{V_L p}{p_L + p}, \tag{3.31}$$

$$V(p) = \frac{V_m C \frac{p}{p_o}}{1 - \frac{p}{p_o}} \left[\frac{1 - (n+1)\left(\frac{p}{p_o}\right)^n + n\left(\frac{p}{p_o}\right)^{n+1}}{1 + (C-1)\frac{p}{p_o} - C\left(\frac{p}{p_o}\right)^{n+1}} \right]. \quad (3.35)$$

The conductivity of fracture network is sensitive to the changes of stress and strain during production due to stress corrosion affecting the proppant strength, crushing, and embedment into the formation (Ghosh, Rai, Sondergeld, & Larese, 2014). To simulate these geomechanic effects during production, displacement equation and flow conservation equation should be coupled as shown in Eqs. (3.98) and (3.99), respectively:

$$\nabla \cdot \left[\mathbf{C} : \frac{1}{2}\left(\nabla \mathbf{u} + (\nabla \mathbf{u})^T\right) \right] + \nabla \cdot [(\alpha p - \eta \Delta T)\mathbf{I}] = \rho_r \mathbf{B}. \quad (3.98)$$

$$\frac{\partial}{\partial t}(\phi \rho_f) - \nabla \cdot \left[\rho_f \frac{\mathbf{k}}{\mu}(\nabla p - \rho_f \mathbf{b}) \right] = Q_f. \quad (3.99)$$

Using the iterative coupled method, porosity can be computed by Eq. (3.100), which is a function of pressure, temperature, and total mean stress:

$$\phi_{n+1} = \phi_n + (c_0 + c_2 a_1)(p - p^n) + (c_1 + c_2 a_2)(T - T^n). \quad (3.100)$$

Permeability can be calculated by empirical correlations. Exponential and power law relationships (Eqs. 3.107 and 3.109) are used to model the variation of permeability depending on stress change during production as shown:

$$k = k_i e^{-b\left(\sigma' - \sigma_i'\right)}, \quad (3.107)$$

$$k = k_i \left(\frac{\sigma'}{\sigma_i'}\right)^{-d}. \quad (3.109)$$

To consider a change of phase behavior in nanopores of shale reservoirs, confinement effects should be modeled. As shown in Chapter 3, these confinement effects can be depicted by critical point shift correlations. The critical pressure and temperature are shifted depending on the pore throat radius as shown in Eqs. (3.112) and (3.113):

$$\Delta T_c^* = \frac{T_{cb} - T_{cp}}{T_{cb}} = 0.9409\frac{\sigma_{LJ}}{r_p} - 0.2415\left(\frac{\sigma_{LJ}}{r_p}\right)^2, \quad (3.112)$$

$$\Delta p_c^* = \frac{p_{cb} - p_{cp}}{p_{cb}} = 0.9409\frac{\sigma_{LJ}}{r_p} - 0.2415\left(\frac{\sigma_{LJ}}{r_p}\right)^2. \quad (3.113)$$

In the development of shale resources, there are high cost and significant uncertainties due to numerous inestimable and uncertain parameters such as reservoir permeability, porosity, hydraulic and natural fracture properties, gas adsorption/desorption, geomechanic properties, confinement effects, and well spacing. Therefore, approaches for sensitivity analysis, history matching, optimization, and uncertainty assessment for the economic development of shale resources is clearly desirable (CMG, 2017a; Yu, 2015). An overall reservoir simulation framework to perform sensitivity analysis, history matching, optimization, and uncertainty assessment with the simple introduction of several systematic and statistic approaches for shale reservoirs are presented in the following section.

Sensitivity analyses are used to determine how much of an effect parameters have on the simulation results and to identify which parameters have the most significant impact on objective functions such as history match error. The main parameters are determined and used for matching variables of history matching, and the other parameters are kept at their estimated values. In a sensitivity analysis, a smaller number of simulation runs are performed to determine the sensitive parameters. This information is used to design history matching or optimization studies, which require a more significant number of simulation runs. There are several methods for sensitivity analysis.

One-parameter-at-a-time (OPAAT) sampling is a traditional method for sensitivity analysis. In this method, each parameter is analyzed independently, whereas other parameters are fixed to their reference values. This procedure is repeated in turn for all parameters to be studied. It measures the effect of each parameter on the objective function while removing the effects of the other parameters. Although OPAAT is simple to use and results are easy to understand without complication by effects of different parameters, the use of OPAAT sampling is generally discouraged by researchers because of several reasons: Results are focused around the reference values, results can vary dramatically if reference values change, more runs are required for the same estimation, parameter interactions cannot be obtained, and conclusively, results from the analysis are not general.

Response surface methodology (Box and Draper, 1987, 2007; Box & Wilson, 1951) presents the correlations between input variables and responses, or objective functions. In contrast with OPAAT, multiple parameters are adjusted together, and then results are analyzed by fitting a polynomial equation, which is called response surface (Fig. 4.10). In Response Surface Methodology, a set of designed experiments is used to build a proxy model to represent the original complicated reservoir simulation model. The response surface is a proxy of the reservoir simulator for fast estimation of the response. The most common proxy models are

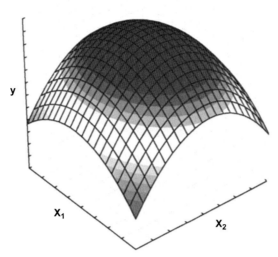

FIG. 4.10 Response surface methodology. (Credit: Bezerra, M. A., Santelli, R. E., Oliveira, E. P., Villar, L. S., & Escaleira, L. A. (2008). Response surface methodology (RSM) as a tool for optimization in analytical chemistry. *Talanta, 76*(5):965–977. https://doi.org/10.1016/j.talanta.2008.05.019.)

linear, simple quadratic, and quadratic form of a polynomial function as follows:

$$y = a_0 + a_1x_1 + a_2x_2 + \ldots + a_kx_k, \qquad (4.29)$$

$$y = a_0 + \sum_{i=1}^{k} a_ix_i + \sum_{i=1}^{k} a_{ii}x_i^2, \qquad (4.30)$$

$$y = a_0 + \sum_{i=1}^{k} a_ix_i + \sum_{i=1}^{k} a_{ii}x_i^2 + \sum_{i<j} \sum_{j=2}^{k} a_{ij}x_ix_j, \qquad (4.31)$$

where y is the response, or the objective function, a_k the coefficients of linear terms, x_k the input variables, a_{ii} the coefficients of quadratic terms, and a_{ij} the coefficients of interaction terms. If parameters have a nonlinear relationship with the objective functions, a quadratic term may be given. If modifying two parameters at the same time has a stronger effect than the sum of individual linear or quadratic effect, an interaction term may also be given. Central composite design (CCD) and D-optimal design are also popular and often used to fit the second-degree surface model (Myers, Montgomery, & Anderson-Cook, 2016). After a proxy model is generated, tornado plots that display a sequence of parameter estimates can be used to analyze the sensitivity of parameters.

Reservoir response is often nonlinear or could be dependent on multiple parameters. Therefore, it is difficult to model with simple trends or polynomial equations. Sobol and Morris methods provide complex relationships in a simplified manner. The Sobol method (1993) is a type of variance-based sensitivity analysis. The main idea of variance-based methods is to quantify the amount of variance that each input factor contributes to the unconditional variance of output. The Morris method (1991) of global sensitivity analysis, which is also called the elementary effects method, is a screening method used to identify model inputs that have the most considerable influence on model outputs. It is based on a replicated, randomized one-at-a-time (OAT) experiment design where only one input parameter is assigned a new value in each run.

History-matching technique is a process to match simulation results with production history data. To obtain a better understanding of reservoir parameters and to simulate the accurate shale model for confidence in forecasting results, history-matching technique is performed. A base numerical shale model created by estimated reservoir properties may present uncertainty. It is adjusted with parameters obtained from sensitivity analysis. Based on the range of values from sensitivity analysis, extensive experiments are performed. As simulation experiment completes, the matching error between a model and production data is calculated. Then, the parameter values for the new simulation job is determined by an optimization method. As more simulation jobs are completed, the results converge to optimal solutions, which provide a satisfactory match with the lowest error between the simulation results and field production data. Various optimization approaches have been used in history matching of reservoir simulation, including Random Search, Differential Evolution (DE), Particle Swarm Optimization (PSO), and Designed Evolution and Controlled Exploration (DECE) (CMG, 2017b; Kennedy & Eberhart, 1995; Storn & Price, 1995; Yang, Card, & Nghiem, 2009, 2007).

Random search methods are the most straightforward stochastic optimization, and they are quite useful in some problems such as small search space and fast-running simulation (CMG, 2017a). There are many different algorithms for random search such as blind random search, localized random search, and enhanced localized random search. The blind random search is the simplest random search method, where the current sampling does not take into account the previous samples. That is, this blind search approach does not adopt the current sampling strategy to information that has been garnered in the search process. One advantage of blind random search is that it is guaranteed to converge to the optimum solution as the number of function evaluations (simulations) gets large. Realistically, however, this convergence feature may have limited use in practice because the algorithm may take a prohibitively large number of function evaluations (simulations) to reach the optimum solution.

Differential Evolution, which is a population-based stochastic optimization technique, was introduced by Storn and Price (1995). The system is initialized with a population of random solutions. It searches for optima by updating this population through three steps: mutation, crossover, and selection. The mutation process involves adding a scaled difference of two solutions to the best solution in each population, to generate a new population. The crossover operation increases the newly generated population diversity. Finally, a selection operator is applied to preserve the optimal solutions for the next generation. The results of the new population are compared against the old population. Each experiment is examined, and the better experiment is kept.

Particle Swarm Optimization is a population-based stochastic optimization technique developed by Kennedy and Eberhart (1995), inspired by social behavior of bird flocking and fish schooling. Social influence and social learning enable a person to maintain cognitive consistency. People solve problems by talking with other people about them and, as they interact, their beliefs, attitudes, and behaviors change. The changes can be depicted as the individuals moving toward one another in a sociocognitive space. Particle swarm simulates this kind of social optimization. The system is initialized with a population of random solutions and searches for optima by updating generations. The individuals iteratively evaluate their candidate solutions and remember the location of their best success so far, making this information available to their neighbors. They are also able to see where their neighbors have had success. Movements through the search space are guided by these successes, with the population usually converging toward good solutions.

DECE optimization method (CMG, 2017a; Yang et al., 2009, 2007) is used to adjust parameter values so that the history match error is minimized. The DECE optimization method is based on the process, which reservoir engineers commonly use to solve history matching or optimization problems. For simplicity, DECE optimization can be described as an iterative optimization process that applies a designed exploration stage first and then a controlled evolution stage. In the designed exploration stage, the goal is to explore the search space in a designed random manner such that maximum information about the solution space can be obtained. In this stage, experimental design and Tabu search techniques are applied to select parameter values and create representative simulation data sets. In the controlled evolution stage, statistical analyses are performed for the simulation results obtained in the designed exploration stage. Based on the analyses, the DECE algorithm scrutinizes every candidate value of each parameter to determine if there is a better chance to improve solution quality if specific candidate values are rejected from being picked again. These rejected candidate values are remembered by the algorithm, and they will not be used in the next controlled exploration stage. To minimize the possibility of being trapped in local minima, the DECE algorithm checks rejected candidate values from time to time to make sure previous rejection decisions are still valid. If the algorithm determines that certain rejection decisions are not valid, the rejection decisions are recalled, and corresponding candidate values are used again.

Optimization studies are used to produce optimum field development plans and operating conditions.

Optimization and history-matching presents a similar methodology that finds either a maximum or minimum value for objective functions. Optimization presents the most improved objective functions such as hydrocarbon recovery and NPV, whereas history matching provides results with a minimum error between simulation models and production data. In shale reservoirs, the optimization of hydraulic fracture parameters and well spacing are significantly important to obtain economic completion scenario. There have been a significant number of attempts in recent years to optimize the design of transverse fractures of horizontal wells for shale gas reservoirs (Bagherian et al., 2010; Bhattacharya & Nikolaou, 2011; Britt & Smith, 2009; Gorucu & Ertekin, 2011; Marongiu-Porcu, Wang, & Economides, 2009; Meyer, Bazan, Henry Jacot, & Lattibeaudiere, 2010; Yu, 2015; Zhang, Du, Deimbacher, Martin, & Harikesavanallur, 2009).

After history matching and optimization, there may be still remaining uncertainties in the value of some reservoir variables. Once several acceptable history-matching models are obtained, the uncertainty of each model should be assessed. In the forecasting models, results are evaluated to confirm the impact of uncertain parameters. A more rigorous assessment of the finally selected models is performed by adjusting some parameters. Although some parameters have little effect on the history-matching process, they may be expected to have a higher impact in the forecast process. A response surface was constructed by simulating a set of cases with parameters that were varied within the range of uncertainty. The response surface was then used as a proxy model that was later used in the Monte Carlo simulation. Once the Monte Carlo simulation was completed, an expected range and probability distribution of results were obtained.

In the following sections, field simulation studies for shale gas and oil reservoirs are presented. Kim (2018) modeled realistic shale gas and oil reservoirs considering various advanced mechanisms mentioned earlier. Barnett and Marcellus Shale gas reservoirs and East Texas shale oil reservoir were modeled. For numerical modeling, compositional and unconventional simulator CMG-GEM (CMG, 2017b) was used. Because of the uncertainty of information obtained from the shale reservoir, model validation is performed with a history-matching technique. By validating the models with history data, information on uncertainties of shale reservoir is obtained. History matching was performed with field production data in each reservoir considering mono- and multilayer adsorption/desorption, stress-dependent compaction, and confinement effects as well as geologic reservoir and fracture properties. History matching was performed with CMG-CMOST (CMG, 2017a). In each reservoir, production forecast was performed to evaluate the effects of mechanisms.

FIELD APPLICATION OF SHALE GAS RESERVOIRS

To verify the effects of various transport mechanisms in shale gas reservoirs, Kim (2018) constructed the numerical model of Barnett Shale. The Barnett Shale is located in the Fort Worth Basin of northern Texas. The reservoir rock itself is a Mississippian age organic-rich shale. Barnett is noted for having generally high silica content (35%–50%), relatively low clay content (<35%), and significant organic carbon content (3%–10%) (Montgomery, Jarvie, Bowker, & Pollastro, 2005). As reported by Bowker (2007), Jarvie, Hill, Ruble, and Pollastro (2007), and Loucks and Ruppel (2007), the composition of Barnett Shale is 40%–45% quartz, 27%–40% clay, and variable contents of pyrite, feldspar, calcite, dolomite, and phosphate.

Fig. 4.11 shows the daily pressure and gas production data reproduced from Anderson et al. (2010). Almost 600 days of production data were available to perform history matching and to evaluate the well performance. The shale gas reservoir with a volume of $3250 \times 550 \times 100 \, \text{ft}^3$ with no flow outer boundaries was constructed, and the horizontal well was drilled whole 3250 ft laterally. To depict natural fractures and a matrix system of a shale reservoir, a dual permeability model was applied. Hydraulic fractures are modeled with a LGR technique to formulate thin blocks assigned with the properties of hydraulic fractures (Rubin, 2010). Within the LGR, cells increase in size logarithmically away from the fracture. Fluids include gas and water, but water is immobile so that the flowing fluid is assumed to be single-phase gas.

Mechanical properties of Barnett Shale were investigated by Vermylen (2011) based on well logs, lab analyses from the sample, and reasonable estimates from the gas shale and rock mechanics literature. Depending on their study, 0.23 and 40 GPa of Poisson's ratio and Young's modulus were used in the Barnett Shale model. Cho, Ozkan, and Apaydin (2013) presented the coefficient of exponential correlation for this Barnett Shale data as 0.0087. Because there are no data for the coefficient of power law correlation, it was calculated by the relationship between exponential and power law correlations. Using the correlation obtained from Dong et al. (2010), a power law correlation coefficient of Barnett Shale was determined as 0.31. In addition, for history

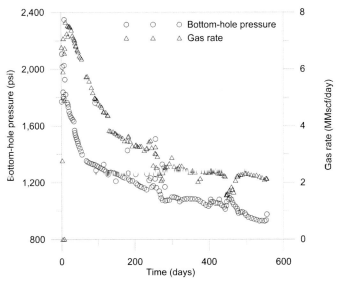

FIG. 4.11 Pressure and gas production rate of Barnett Shale. (Credit: Reproduced from Anderson, D. M., Nobakht, M., Moghadam, S., & Mattar, L. (2010). Analysis of production data from fractured shale gas wells. *Paper presented at the SPE unconventional gas Conference, Pittsburgh, Pennsylvania, USA.* https://doi. org/10.2118/131787-MS.)

matching, the range of exponential and power law compaction coefficients is determined based on the experiments of Dong et al. (2010). Fig. 4.12 provides a permeability multiplier based on these coefficients for Barnett Shale simulation.

Table 4.2 provides the parameters and their ranges used in the history matching of Barnett Shale reservoir. Because of the limited data availability, the significant uncertainty lies in the reservoir and fracture properties. Fig. 4.13 shows the best-matched models for the well bottom-hole pressure in three different cases. In this study, cases 1, 2, and 3 indicate the noncompaction, the compaction with exponential correlation, and the compaction with power law correlation models. The best-matched models show 6.7%, 5.8%, and 6.4% of matching error, respectively. Models with exponential and power law correlations show lower matching errors than the noncompaction model. Although the model with exponential correlation shows the lowest matching error, it cannot be determined directly as the exact model. In applications of shale reservoir history matching, nonuniqueness problem is significant because of the limitation of reservoir information. Tables 4.3–4.5 provide the best-matched parameters and ranges of top 25 models after history matching in three different compaction cases. As shown in the tables, best-matched values and ranges are various in each case. Especially, hydraulic fracture half-length and spacing

present significantly different in three compaction cases. Depending on RTA results of Anderson et al. (2010), hydraulic fracture half-length and spacing are 250 and 174 ft, respectively. As shown in Table 3.4, models with power law correlation correspond well with RTA results.

The production forecast was conducted for another 10 years using the selected top 25 models. The prediction run was constrained with the last wellhead pressure. Fig. 4.14 shows the predicted cumulative production of gas for three compaction cases. Although the small difference is observed in history-matching results, the effect of it increases extremely in future prediction. Ranges of cumulative gas production of models with noncompaction, compaction with exponential correlation, and compaction with power law correlation at the end of the simulation are 4.5–5.6 Bscf, 2.3–2.9 Bscf, and 3.2–4.2 Bscf, respectively. Because of these significant variations in production forecast process, stress-dependent compaction should be considered to model the shale gas reservoir.

Marcellus Shale was also considered to evaluate the effects of transport mechanisms. The Marcellus Shale is located in the Appalachian Basin across six states, including Pennsylvania, New York, West Virginia, Ohio, Virginia, and Maryland. Although there are many horizontal wells drilled along with multistage hydraulic fracturing in the Marcellus Shale, the

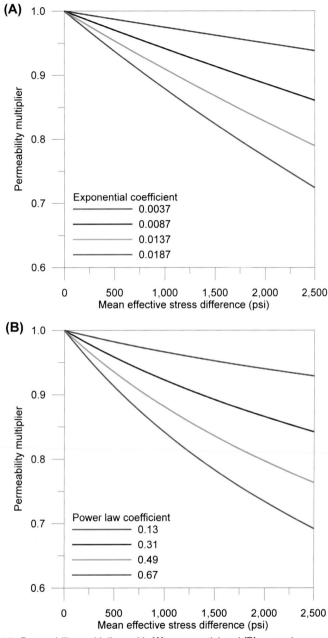

FIG. 4.12 Permeability multipliers with **(A)** exponential and **(B)** power law correlations.

completion effectiveness is not entirely understood. Marcellus is generally composed of high silica content, high clay content, and moderate organic carbon content (3%) (Lora, 2015). Lora (2015) presented quantitative analysis for Marcellus Shale: 21% quartz, 69% clay, and variable contents of gypsum, pyrite, and calcite.

One-hundred and eighty days of flowing bottom-hole pressure and gas production rate of Marcellus reproduced from the study of Yu (2015) were used to perform history matching. The reservoir model was constructed with dimension of 3105 ft×1000 × 137 ft, which corresponds to length, width, and thickness,

TABLE 4.2
Parameter Uncertainties for History Matching of Barnett Shale Gas Reservoir

Uncertain Parameters	Base	Low	High
Initial reservoir pressure (psi)	2,100	1,800	2,400
Hydraulic fracture permeability (md)	100,000	10,000	100,000
Matrix permeability (md)	0.1×10^{-5}	0.1×10^{-5}	10×10^{-5}
Natural fracture permeability (md)	0.20	0.01	0.50
Natural fracture spacing (ft)	50	25	75
Matrix porosity	0.36	0.02	0.1
Natural fracture porosity	0.08	0.00002	0.1
Hydraulic fracture half-length (ft)	100	50	400
Hydraulic fracture spacing (ft)	174	100	400
Exponential coefficient	0.0087	0.0037	0.0187
Power law coefficient	0.31	0.13	0.67

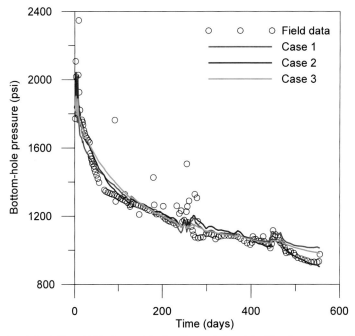

FIG. 4.13 Best-matched Barnett Shale models for Cases 1 (model without compaction), 2 (model with exponential correlation), and 3 (model with power law correlation).

respectively. The reservoir has two different shale layers. Porosity of the top layer is 9%, and porosity of the bottom layer is around 13.8%. Thicknesses of the top and bottom layers are 94 and 43 ft, respectively. The well was drilled in the bottom layer and completed using a lateral length of 2605 ft. Even though Yu (2015) performed history matching with these well data, he only considered hydraulic fracture properties and matrix permeability so that matching results are inadequate. In this model, a dual permeability model was applied

TABLE 4.3
Parameters for Top 25 Matched Models of Barnett Shale Gas Reservoir: Case 1

Uncertain Parameters	Best	Low	High
Initial reservoir pressure (psi)	1,922	1,895	2,053
Hydraulic fracture permeability (md)	86,500	75,700	98,650
Matrix permeability (md)	5.6×10^{-5}	2.5×10^{-5}	5.9×10^{-5}
Natural fracture permeability (md)	0.3824	0.3236	0.4853
Natural fracture spacing (ft)	66	60	74
Matrix porosity	0.090	0.064	0.098
Natural fracture porosity	0.0245	0.0230	0.0365
Hydraulic fracture half-length (ft)	204	204	365
Hydraulic fracture spacing (ft)	271	271	348

TABLE 4.4
Parameters for Top 25 Matched Models of Barnett Shale Gas Reservoir: Case 2

Uncertain Parameters	Best	Low	High
Initial reservoir pressure (psi)	2,048	1,922	2,132
Hydraulic fracture permeability (md)	98,200	77,500	98,650
Matrix permeability (md)	5.7×10^{-5}	5.6×10^{-5}	8.7×10^{-5}
Natural fracture permeability (md)	0.206	0.169	0.341
Natural fracture spacing (ft)	42	32	49
Matrix porosity	0.030	0.029	0.042
Natural fracture porosity	0.0225	0.0050	0.0240
Hydraulic fracture half-length (ft)	99	71	171
Hydraulic fracture spacing (ft)	301	208	316
Exponential coefficient	0.0087	0.0037	0.0137

TABLE 4.5
Parameters for Top 25 Matched Models of Barnett Shale Gas Reservoir: Case 3

Uncertain Parameters	Best	Low	High
Initial reservoir pressure (psi)	1,859	1,801	1,922
Hydraulic fracture permeability (md)	96,400	85,150	100,000
Matrix permeability (md)	5.4×10^{-5}	5.4×10^{-5}	9.8×10^{-5}
Natural fracture permeability (md)	0.186	0.186	0.336
Natural fracture spacing (ft)	73	68	75
Matrix porosity	0.062	0.035	0.070
Natural fracture porosity	0.0295	0.0135	0.0295
Hydraulic fracture half-length (ft)	236	159	264
Hydraulic fracture spacing (ft)	196	102	196
Power law coefficient	0.31	0.13	0.49

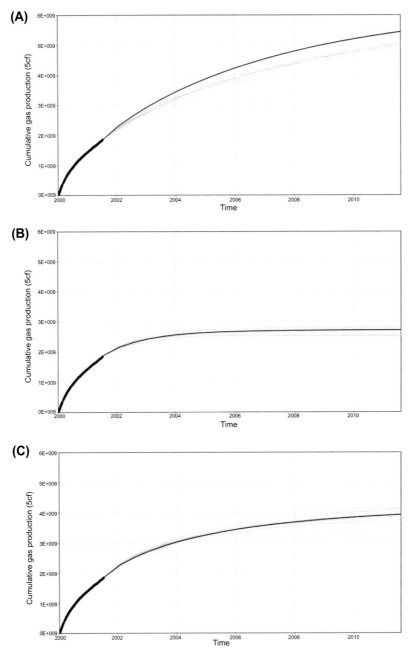

FIG. 4.14 Barnett production forecast of Cases **(A)** 1, **(B)** 2, and **(C)** 3 for 10 years.

to consider the natural fracture system. LGR technique with increasing cells in size logarithmically away from the fracture is applied to model hydraulic fractures. Single-phase gas flow is assumed owing to immobile water in shale rock.

Adsorption data presented by Yu (2015) were used to model the Marcellus Shale. Yu (2015) presented experimental measurements of CH_4 adsorption from the Marcellus Shale core samples that deviate from the Langmuir isotherm but obey the BET isotherm. Both

TABLE 4.6
Parameter Uncertainties for History Matching of Marcellus Shale Gas Reservoir

Uncertain Parameters	Base	Low	High
Initial reservoir pressure (psi)	4,300	3,800	4,800
Hydraulic fracture permeability (md)	10,000	5,000	60,000
Matrix permeability (md)	8.0×10^{-4}	1.0×10^{-4}	10×10^{-4}
Natural fracture spacing (ft)	50	1	50
Hydraulic fracture half-length (ft)	300	50	500
Hydraulic fracture spacing (ft)	50	50	150
Exponential coefficient	0.0087	0.0037	0.0187
Power law coefficient	0.31	0.13	0.67

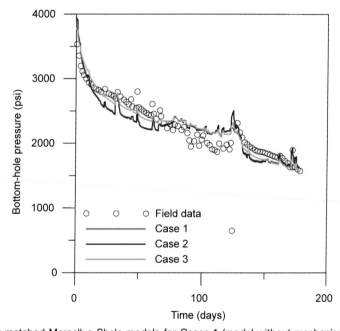

FIG. 4.15 Best matched Marcellus Shale models for Cases 1 (model without mechanism), 2 (model with multilayer adsorption and confinement effects), and 3 (model with compaction, multilayer adsorption, and confinement effects).

Langmuir and BET isotherms matched five adsorption data, and they were used for history-matching parameters. Stress-dependent compaction is also considered in this model. Because Marcellus Shale has similar rock characteristics with Barnett Shale, the same ranges of compaction coefficients were used in this model. Confinement effect is also considered. Pore throat radius is calculated by Eq. (4.32) (Al Hinai, Rezaee,

Ali, & Roland, 2013), and then critical pressure and temperature are calculated by Eqs. (3.112)–(3.114). In this history-matching process, critical points are automatically calculated by reservoir properties. Table 4.6 gives the parameters and their ranges used in the history matching of Marcellus Shale reservoir.

$$\log k = 37.255 - 6.345 \log \phi + 15.227 \log r_p. \quad (4.32)$$

TABLE 4.7
Parameters for Top 25 Matched Models of Marcellus Shale Gas Reservoir: Case 1

Uncertain Parameters	Best	Low	High
Initial reservoir pressure (psi)	3,985	3,905	4,300
Hydraulic fracture permeability (md)	59,725	10,000	60,000
Matrix permeability (md)	9.6×10^{-4}	8.0×10^{-4}	9.9×10^{-4}
Natural fracture spacing (ft)	43	34	50
Hydraulic fracture half-length (ft)	320	300	365
Hydraulic fracture spacing (ft)	97	50	113

TABLE 4.8
Parameters for Top 25 Matched Models of Marcellus Shale Gas Reservoir: Case 2

Uncertain Parameters	Best	Low	High
Initial reservoir pressure (psi)	3,965	3,915	4,300
Hydraulic fracture permeability (md)	59,725	10,000	60,000
Matrix permeability (md)	9.6×10^{-4}	8.0×10^{-4}	1.0×10^{-3}
Natural fracture spacing (ft)	46	40	50
Hydraulic fracture half-length (ft)	322	300	338
Hydraulic fracture spacing (ft)	97	50	100

TABLE 4.9
Parameters for Top 25 Matched Models of Marcellus Shale Gas Reservoir: Case 3

Uncertain Parameters	Best	Low	High
Initial reservoir pressure (psi)	3,800	3,800	4,800
Hydraulic fracture permeability (md)	37,175	10,000	50,925
Matrix permeability (md)	6.5×10^{-4}	7.4×10^{-4}	8.2×10^{-4}
Natural fracture spacing (ft)	14	6	50
Hydraulic fracture half-length (ft)	223	190	300
Hydraulic fracture spacing (ft)	50	50	66
Power law coefficient	0.31	0.13	0.67

Fig. 4.15 shows history-matching results of Marcellus Shale. In this graph, Case 1 indicates model without advanced mechanisms suggested in this study, Case 2 indicates model with multilayer adsorption and confinement effects, and Case 3 indicates model that considers multilayer adsorption, confinement effect, and stress-dependent compaction. As shown in Fig. 4.15, best-matched models of Cases 1, 2, and 3 show 4.7%, 4.7%, and 4.0% of matching error, respectively. Tables 4.7—4.9 provide the best-matched parameters and ranges of top 25 models after history matching in each case. As shown in the matching results, Cases 1

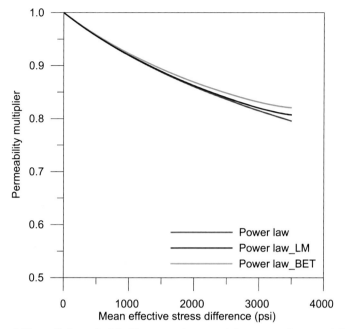

FIG. 4.16 Permeability multiplier calculated by power law correlation, power law correlation coupled with Langmuir isotherm, and power law correlation coupled with BET isotherm.

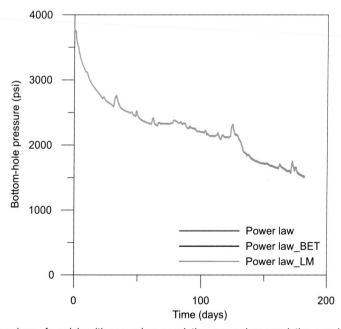

FIG. 4.17 Comparison of models with power law correlation, power law correlation coupled with Langmuir Isotherm, and power law correlation coupled with BET isotherm.

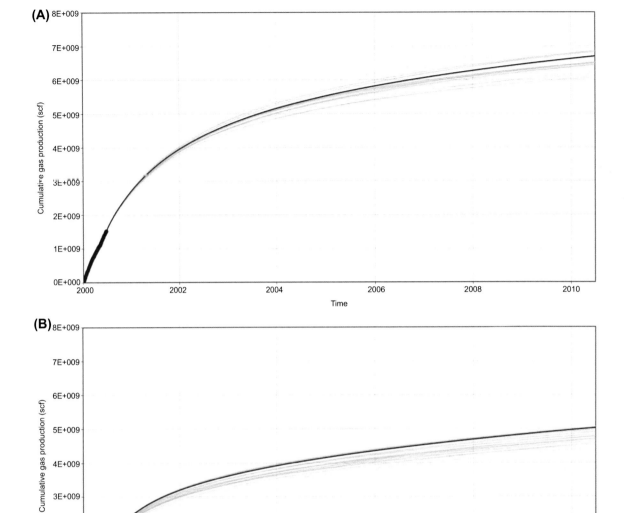

FIG. 4.18 Marcellus production forecast of Cases **(A)** 1 and **(B)** 3 for 10 years.

and 2 present similar models. According to results of Cases 1 and 2, effects of multilayer adsorption and confinement are insignificant in Marcellus Shale. Because of low initial reservoir pressure, Langmuir and BET isotherms present similar behavior during primary production, and confinement effect is also insignificant in the dry gas reservoir. Although confinement effects change viscosity and original gas in place (OGIP) by phase behavior shift, the influence is small compared with compaction effects. In this model, initial gas viscosity decreases from 0.0255 to 0.0221 cp, and OGIP decreases from 2430 to

2407 MMscf. However, as shown in Case 3, stress-dependent compaction is important in Marcellus Shale reservoir.

In the top 25 models, 23 models are matched with power law correlation, and power law coefficients of 21 models are higher than 0.31. Similar with Barnett Shale, power law correlation presents better matching results than exponential correlation. In addition, matched power law coefficients of Marcellus Shale are slightly higher than that of Barnett Shale. Depending on Vermylen (2011), a large amount of clay suggests that this rock is mechanically weak and it could deform plastically and creep during the long periods under stress. Because Marcellus Shale generally contains a more substantial amount of clay than Barnett Shale (Lora, 2015; Montgomery et al., 2005), results of history matching are reasonable. Fig. 4.16 shows permeability multiplier calculated by power law correlation, power law correlation coupled with Langmuir isotherm, and power law correlation combined with BET isotherm. Permeability change by adsorption effects is small compared with stress effects. Permeability changes caused by both Langmuir and BET isotherms are under 1% when mean effective stress changes by 2500 psi and permeability changes 17% by stress effects. Fig. 4.17 shows bottom-hole pressure results to investigate the impact of adsorption on compaction. In the condition of reservoir production, effects of adsorption on permeability are negligible compared with stress-dependent compaction.

The production forecast was also conducted in Marcellus Shale for another 10 years using the selected top 25 models. The prediction run was constrained with the last wellhead pressure. Forecast of cumulative gas production from existing and improved models are shown in Fig. 4.18. Ranges of gas production at the end of prediction are 6.0–7.2 Bscf and 4.5–5.1 Bscf in Cases 1 and 3, respectively. Case 2 shows similar results with Case 1. Even if history-matching results show only 0.7% difference between Cases 1 and 3, 10-year prediction presents about 40% difference between them. In history matching of Barnett and Marcellus Shales, which have similar characteristics of rock, stress-dependent compaction is an important mechanism, and power law correlation presents competent simulation output.

FIELD APPLICATION OF SHALE OIL RESERVOIRS
The shale oil reservoir model of East Texas was constructed to analyze complex transport mechanisms. Well bottom-hole pressure, oil rate, and water rate for history matching were obtained from Iino et al. (2017a, 2017b). The well opened at 1300 bbl/d liquid rate and declined to 100 bbl/d during the observation period. High water cut during the early production was due to the recovery of completion fluid. The modeled reservoir section is dimensioned 7100 ft×2500 × 180 ft. The reservoir is undersaturated at the initial pressure of 3953 psi against the bubble point pressure of 2930 psi. A dual permeability model is assumed to consider natural fracture system. Iino et al. (2017a) presented matrix properties derived from the core and log interpretations. Porosity, permeability, and initial water saturation are 0.08, 2.7×10^{-5} md, and 0.41, respectively. The connate water saturation was set to be 0.29.

TABLE 4.10
Parameter Uncertainties for History Matching of East Texas Shale Oil Reservoir

Uncertain Parameters	Base	Low	High
Hydraulic fracture permeability (md)	50,000	1,000	60,000
Matrix permeability (md)	0.001	0.00001	0.001
Natural fracture permeability	0.000027	0.00002	0.01
Matrix porosity	0.067	0.05	0.08
Natural fracture porosity	0.005	0.003	0.008
Natural fracture spacing (ft)	100	1	100
Hydraulic fracture half-length (ft)	200	100	400
Hydraulic fracture spacing (ft)	200	100	500
Exponential coefficient	0.005	0.0005	0.001

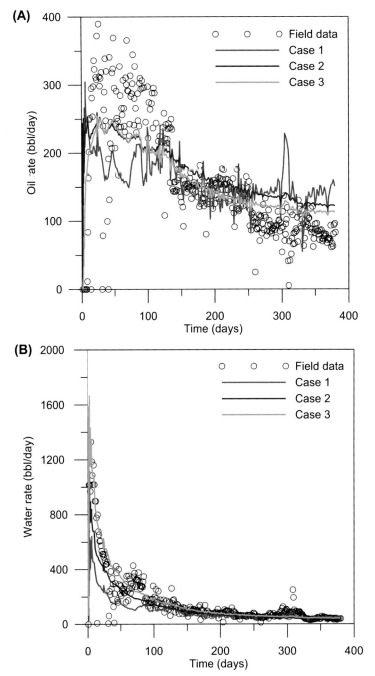

FIG. 4.19 Best-matched **(A)** oil rate and **(B)** water rate of East Texas shale models for Cases 1 (model without mechanism), 2 (model with compaction), and 3 (model with compaction and confinement effects).

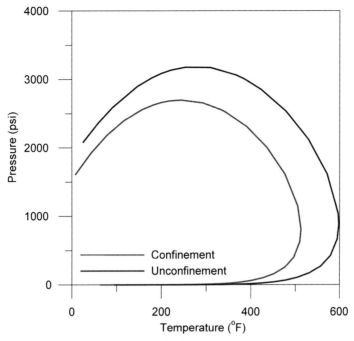

FIG. 4.20 Phase envelope with and without confinement effects.

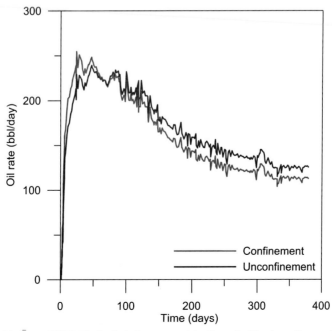

FIG. 4.21 Comparison of oil rate between model with and without confinement effects.

Table 4.10 shows reservoir and fracture parameters and their ranges used in the history matching of East Texas shale oil reservoir. Because of the limited data availability, the significant uncertainty lies in the reservoir and fracture properties. With these matching parameters, three different cases were used for history matching. In this study, model without advanced mechanisms (Case 1), model with compaction (Case 2), and model with compaction and confinement effects (Case 3) were compared. Fig. 4.19 shows the best matched models for the oil and water rates in three different cases. As shown in Fig. 4.19, Case 1 cannot match both oil and water rates. Matching errors decrease with the consideration of stress-dependent compaction and confinement effect. Best matched models of Cases 1, 2, and 3 present 13.9%, 10.4%, and 9.9%, respectively, of matching errors with field history data. To consider confinement effects, critical pressure and temperature of each component were calculated by using Eqs. (3.112)−(3.114). Pore throat radius of shale reservoir is calculated by using Eq. (4.32), and critical points are computed in every model of history matching. Fig. 4.20 shows phase behavior depending on the consideration of confinement effect. In East Texas shale

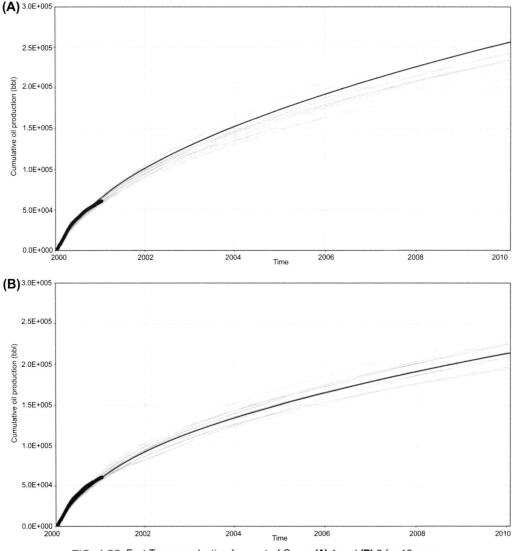

FIG. 4.22 East Texas production forecast of Cases **(A)** 1 and **(B)** 2 for 10 years.

model, oil viscosity decreases from 0.1242 to 0.0748 cp and original oil in place (OOIP) decreases from 6766 to 3738 Mstb. Effects of phase behavior shift in East Texas shale are presented in Fig. 4.21. The model considering confinement effects and the model not considering confinement effects were presented. Because of the decrease of oil viscosity by phase behavior shift, the productivity of early stage is higher than the model without confinement effect. However, the oil rate of the model with confinement effects decreases because of smaller OOIP due to density shift calculated by critical point change.

The oil production forecast was also conducted in East Texas shale for another 10 years using the selected top 25 models. The prediction run was constrained with the last wellhead pressure. Fig. 4.22 shows cumulative oil production forecast of the model with compaction and model with compaction and confinement effects. Ranges of oil production at the end of the forecast are 196–288 Mbbl and 170–263 Mbbl in each model. In East Texas shale oil model, stress-dependent compaction and confinement effects should be considered in detail.

REFERENCES

Adachi, J., Siebrits, E., Peirce, A., & Desroches, J. (2007). Computer simulation of hydraulic fractures. *International Journal of Rock Mechanics and Mining Sciences, 44*(5), 739–757. https://doi.org/10.1016/j.ijrmms.2006.11.006.

Advani, S. H., Khattab, H., & Lee, J. K. (1985). Hydraulic fracture geometry modeling, prediction, and comparisons. In *Paper presented at the SPE/DOE low permeability gas reservoirs symposium, Denver, Colorado.* https://doi.org/10.2118/13863-MS.

Ahmed, U., & Meehan, D. N. (2016). *Unconventional oil and gas resources: Exploitation and development.* CRC Press.

Al Hinai, A., Rezaee, R., Ali, S., & Roland, L. (2013). Permeability prediction from mercury injection capillary pressure: An example from the Perth basin, Western Australia. *The APPEA Journal, 53*(1), 31–36.

Al-Kobaisi, M., Ozkan, E., & Kazemi, H. (2006). A hybrid numerical/analytical model of a finite-conductivity vertical fracture intercepted by a horizontal well. *SPE Reservoir Evaluation & Engineering, 9*(04), 345–355. https://doi.org/10.2118/92040-PA.

Anderson, D. M., Nobakht, M., Moghadam, S., & Mattar, L. (2010). Analysis of production data from fractured shale gas wells. In *Paper presented at the SPE unconventional gas conference, Pittsburgh, Pennsylvania, USA.* https://doi.org/10.2118/131787-MS.

Aybar, U. (2014). *Investigation of analytical models incorporating geomechanical effects on production perdormance of hydraulically and naturally fractured unconventional reservoirs.* Master of Science in Engineering. The University of Texas at Austin.

Aybar, U., Yu, W., Eshkalak, M. O., Sepehrnoori, K., & Patzek, T. (2015). Evaluation of production losses from unconventional shale reservoirs. *Journal of Natural Gas Science and Engineering, 23,* 509–516. https://doi.org/10.1016/j.jngse.2015.02.030.

Bagherian, B., Ghalambor, A., Sarmadivaleh, M., Rasouli, V., Nabipour, A., & Mahmoudi Eshkaftaki, M. (2010). Optimization of multiple-fractured horizontal tight gas well. In *Paper presented at the SPE international symposium and exhibition on formation damage control, Lafayette, Louisiana, USA.* https://doi.org/10.2118/127899-MS.

Bale, A., Smith, M. B., & Settari, A. (1994). Post-frac productivity calculation for complex reservoir/fracture geometry. In *Paper presented at the european petroleum conference, London, United Kingdom.* https://doi.org/10.2118/28919-MS.

Barenblatt, G. I., Zheltov, I. P., & Kochina, I. N. (1960). Basic concepts in the theory of seepage of homogeneous liquids in fissured rocks [strata]. *Journal of Applied Mathematics and Mechanics, 24*(5), 1286–1303. https://doi.org/10.1016/0021-8928(60)90107-6.

Barree, R. D. (1983). A practical numerical simulator for three-dimensional fracture propagation in heterogeneous media. In *Paper presented at the SPE reservoir simulation symposium, San Francisco, California.* https://doi.org/10.2118/12273-MS.

Barree & Associates. (2015). *GOHFER user manual.* Barree & Associates LLC.

Bennett, C. O., Rodolfo, G. C.-V., Reynolds, A. C., & Raghavan, R. (1985). Approximate solutions for fractured wells producing layered reservoirs. *Society of Petroleum Engineers Journal, 25*(05), 729–742. https://doi.org/10.2118/11599-PA.

Bezerra, M. A., Santelli, R. E., Oliveira, E. P., Villar, L. S., & Escaleira, L. A. (2008). Response surface methodology (RSM) as a tool for optimization in analytical chemistry. *Talanta, 76*(5), 965–977. https://doi.org/10.1016/j.talanta.2008.05.019.

Bhattacharya, S., & Nikolaou, M. (2011). Optimal fracture spacing and stimulation design for horizontal wells in unconventional gas reservoirs. In *Paper presented at the SPE annual technical conference and exhibition, Denver, Colorado, USA.* https://doi.org/10.2118/147622-MS.

Bowker, K. A. (2007). Development of the Barnett shale play, Fort Worth Basin. *AAPG Bulletin, 91*(4), 13.

Box, G. E. P., & Draper, N. R. (1987). *Empirical model-building and response surfaces.* Wiley.

Box, G. E. P., & Draper, N. R. (2007). *Response surfaces, mixtures, and ridge analyses.* Wiley.

Box, G. E. P., & Wilson, K. B. (1951). On the experimental attainment of optimum conditions. *Journal of the Royal Statistical Society. Series B (Methodological), 13*(1), 1–45.

Britt, L. K., & Smith, M. B. (2009). Horizontal well completion, stimulation optimization, and risk mitigation. In *Paper presented at the SPE eastern regional meeting, charleston, West Virginia, USA.* https://doi.org/10.2118/125526-MS.

Brown, M., Ozkan, E., Raghavan, R., & Kazemi, H. (2011). Practical solutions for pressure-transient responses of fractured horizontal wells in unconventional shale reservoirs. *SPE Reservoir Evaluation & Engineering, 14*(06), 663–676. https://doi.org/10.2118/125043-PA.

Camacho-V, R. G., Raghavan, R., & Reynolds, A. C. (1987). Response of wells producing layered reservoirs: Unequal fracture length. *SPE Formation Evaluation, 2*(01), 9−28. https://doi.org/10.2118/12844-PA.

Cho, Y., Ozkan, E., & Apaydin, O. G. (2013). Pressure-dependent natural-fracture permeability in shale and its effect on shale-gas well production. *SPE Reservoir Evaluation & Engineering, 16*(02), 216−228. https://doi.org/10.2118/159801-PA.

Cinco-Ley, H., & Meng, H. Z. (1988). Pressure transient analysis of wells with finite conductivity vertical fractures in double porosity reservoirs. In *Paper presented at the SPE annual technical conference and exhibition, Houston, Texas*. https://doi.org/10.2118/18172-MS.

Cinco-Ley, H., & Samaniego-V, F. (1981). Transient pressure analysis for fractured wells. *Journal of Petroleum Technology, 33*(09), 1749−1766. https://doi.org/10.2118/7490-PA.

Cinco, L., Heber, F., Samaniego, V., & Dominguez A, N. (1978). Transient pressure behavior for a well with a finite-conductivity vertical fracture. *Society of Petroleum Engineers Journal, 18*(04), 253−264. https://doi.org/10.2118/6014-PA.

Cipolla, C. L. (2009). Modeling production and evaluating fracture performance in unconventional gas reservoirs. *Journal of Petroleum Technology, 61*(09), 84−90. https://doi.org/10.2118/118536-JPT.

Cipolla, C. L., Fitzpatrick, T., Williams, M. J., & Ganguly, U. K. (2011). Seismic-to-Simulation for unconventional reservoir development. In *Paper presented at the SPE reservoir characterisation and simulation conference and exhibition, Abu Dhabi, UAE*. https://doi.org/10.2118/146876-MS.

Cipolla, C. L., Lolon, E. P., Erdle, J. C., & Rubin, B. (2010). Reservoir modeling in shale-gas reservoirs. *SPE Reservoir Evaluation & Engineering, 13*(04), 638−653. https://doi.org/10.2118/125530-PA.

Cipolla, C. L., Maxwell, S. C., & Mack, M. G. (2012). Engineering guide to the application of microseismic interpretations. In *Paper presented at the SPE hydraulic fracturing technology conference, The Woodlands, Texas, USA*. https://doi.org/10.2118/152165-MS.

Cipolla, C., & Wallace, J. (2014). Stimulated reservoir volume: A misapplied concept? In *Paper presented at the SPE hydraulic fracturing technology conference, The Woodlands, Texas, USA*. https://doi.org/10.2118/168596-MS.

CMG. (2017a). *CMOST user guide*. Computer Modelling Group.

CMG. (2017b). *GEM user guide*. Computer Modelling Group.

Coats, K. H. (1989). Implicit compositional simulation of single-porosity and dual-porosity reservoirs. In *Paper presented at the SPE symposium on reservoir simulation, Houston, Texas*. https://doi.org/10.2118/18427-MS.

Collins, D. A., Nghiem, L. X., Li, Y. K., & Grabonstotter, J. E. (1992). An efficient approach to adaptive- implicit compositional simulation with an equation of state. *SPE Reservoir Engineering, 7*(02), 259−264. https://doi.org/10.2118/15133-PA.

Daneshy, A. A. (1973). On the design of vertical hydraulic fractures. *Journal of Petroleum Technology, 25*(01), 83−97. https://doi.org/10.2118/3654-PA.

Dong, J.-J., Hsu, J.-Y., Wu, W.-J., Shimamoto, T., Hung, J.-H., Yeh, E.-C., et al. (2010). Stress-dependence of the permeability and porosity of sandstone and shale from TCDP Hole-A. *International Journal of Rock Mechanics and Mining Sciences, 47*(7), 1141−1157. https://doi.org/10.1016/j.ijrmms.2010.06.019.

Dowdle, W. L., & Hyde, P. V. (1977). Well test analysis of hydraulically fractured gas wells. In *Paper presented at the SPE deep drilling and production symposium, Amarillo, Texas*. https://doi.org/10.2118/6437-MS.

Evans, R. D., & Civan, F. (1994). *Characterization of non-Darcy multiphase flow in petroleum bearing formation*. Final report. United States.

Fanchi, J. R., Arnold, K., Mitchell Robert, F., Holstein, E. D., & Warner, H. R. (2007). *Petroleum engineering handbook: Production operations engineering*. Society Of Petroleum Engineers.

Fisher, M. K., Wright, C. A., Davidson, B. M., Goodwin, A. K., Fielder, E. O., Buckler, W. S., et al. (2002). Integrating fracture mapping technologies to optimize stimulations in the Barnett shale. In *Paper presented at the SPE annual technical conference and exhibition, San Antonio, Texas*. https://doi.org/10.2118/77441-MS.

Forchheimer, P. (1901). Water movement through the ground. *Zeitschrift Des Vereines Deutscher Ingenieure, 45*, 1781−1788.

Geertsma, J., & De Klerk, F. (1969). A rapid method of predicting width and extent of hydraulically induced fractures. *Journal of Petroleum Technology, 21*(12), 1571−1581. https://doi.org/10.2118/2458-PA.

Geshelin, B. M., Grabowski, J. W., & Pease, E. C. (1981). Numerical study of transport of injected and reservoir water in fractured reservoirs during steam stimulation. In *Paper presented at the SPE annual technical conference and exhibition, San Antonio, Texas*. https://doi.org/10.2118/10322-MS.

Ghosh, S., Rai, C. S., Sondergeld, C. H., & Larese, R. E. (2014). Experimental investigation of proppant diagenesis. In *Paper presented at the SPE/CSUR unconventional resources conference − Canada, Calgary, Alberta, Canada*. https://doi.org/10.2118/171604-MS.

Gidley, J. L., & Society of Petroleum Engineers. (1989). *Recent advances in hydraulic fracturing*. Henry L. Doherty Memorial Fund of AIME, Society of Petroleum Engineers.

Gorucu, S. E., & Ertekin, T. (2011). Optimization of the design of transverse hydraulic fractures in horizontal wells placed in dual porosity tight gas reservoirs. In *Paper presented at the SPE Middle East unconventional gas conference and exhibition, Muscat, Oman*. https://doi.org/10.2118/142040-MS.

Green, C. A., David Barree, R., & Miskimins, J. L. (2007). Development of a methodology for hydraulic fracturing models in tight, massively stacked, lenticular reservoirs. In *Paper presented at the SPE hydraulic fracturing technology conference, College Station, Texas, USA*. https://doi.org/10.2118/106269-MS.

Gringarten, A. C., Henry, J., Ramey, Jr., & Raghavan, R. (1974). Unsteady-state pressure distributions created by a well with a single infinite-conductivity vertical fracture. *Society of Petroleum Engineers Journal, 14*(04), 347−360. https://doi.org/10.2118/4051-PA.

Gringarten, A. C., Ramey, H. J., Jr., & Raghavan, R. (1975). Applied pressure analysis for fractured wells. *Journal of Petroleum Technology*, *27*(07), 887–892. https://doi.org/10.2118/5496-PA.

Heller, R., & Zoback, M. (2014). Adsorption of methane and carbon dioxide on gas shale and pure mineral samples. *Journal of Unconventional Oil and Gas Resources*, *8*, 14–24. https://doi.org/10.1016/j.juogr.2014.06.001.

Holditch, S. A. (1979). Factors affecting water blocking and gas flow from hydraulically fractured gas wells. *Journal of Petroleum Technology*, *31*(12), 1515–1524. https://doi.org/10.2118/7561-PA.

Howard, G. C., & Fast, C. R. (1957). Optimum fluid characteristics for fracture extension. In *Paper presented at the drilling and production practice, New York, New York*.

Iino, A., Vyas, A., Huang, J., Datta-Gupta, A., Fujita, Y., Bansal, N., et al. (2017a). Efficient modeling and history matching of shale oil reservoirs using the fast marching method: Field application and validation. In *Paper presented at the SPE western regional meeting, Bakersfield, California*. https://doi.org/10.2118/185719-MS.

Iino, A., Vyas, A., Huang, J., Datta-Gupta, A., Fujita, Y., & Sankaran, S. (2017b). Rapid compositional simulation and history matching of shale oil reservoirs using the fast marching method. In *Paper presented at the SPE/AAPG/SEG unconventional resources technology conference, Austin, Texas, USA*. https://doi.org/10.15530/URTEC-2017-2693139.

Jarvie, D. M., Hill, R. J., Ruble, T. E., & Pollastro, R. M. (2007). Unconventional shale-gas systems: The Mississippian Barnett Shale of north-central Texas as one model for thermogenic shale-gas assessment. *American Association of Petroleum Geologists Bulletin*, *91*(4), 475–499. https://doi.org/10.1306/12190606068.

Ji, L., Settari, A., Orr, D. W., & Sullivan, R. B. (2004). Methods for modelling static fractures in reservoir simulation. In *Paper presented at the Canadian international petroleum conference, Calgary, Alberta*. https://doi.org/10.2118/2004-260.

Kam, P., Nadeem, M., Novlesky, A., Kumar, A., & Omatsone, E. N. (2015). Reservoir characterization and history matching of the horn river shale: An integrated geoscience and reservoir-simulation approach. *Journal of Canadian Petroleum Technology*, *54*(06), 475–488. https://doi.org/10.2118/171611-PA.

Kazemi, H. (1969). Pressure transient analysis of naturally fractured reservoirs with uniform fracture distribution. *Society of Petroleum Engineers Journal*, *9*(04), 451–462. https://doi.org/10.2118/2156-A.

Kazemi, H., Merrill, L. S., Jr., Porterfield, K. L., & Zeman, P. R. (1976). Numerical simulation of water-oil flow in naturally fractured reservoirs. *Society of Petroleum Engineers Journal*, *16*(06), 317–326. https://doi.org/10.2118/5719-PA.

Kennedy, J., & Eberhart, R. (1995). Particle swarm optimization. In *Proceedings of ICNN'95 — international conference on neural networks, November 27– December 1995*.

Kim, T. H. (2018). *Integrative modeling of CO_2 injection for enhancing hydrocarbon recovery and CO_2 storage in shale reservoirs*. Hanyang University.

Langmuir, I. (1918). The adsorption of gases on plane surfaces of glass, mica and platinum. *Journal of the American Chemical Society*, *40*(9), 1361–1403. https://doi.org/10.1021/ja02242a004.

Li, L., & Lee, S. H. (2008). Efficient field-scale simulation of black oil in a naturally fractured reservoir through discrete fracture networks and homogenized media. *SPE Reservoir Evaluation & Engineering*, *11*(04), 750–758. https://doi.org/10.2118/103901-PA.

Lim, K. T., & Aziz, K. (1995). Matrix-fracture transfer shape factors for dual-porosity simulators. *Journal of Petroleum Science and Engineering*, *13*(3), 169–178. https://doi.org/10.1016/0920-4105(95)00010-F.

Lora, R. V. (2015). *Geomechanical characterization of Marcellus shale*. Master of Science. The University of Vermont.

Loucks, R. G., & Ruppel, S. C. (2007). Mississippian Barnett Shale: Lithofacies and depositional setting of a deepwater shale-gas succession in the Fort Worth Basin, Texas. *AAPG Bulletin*, *91*(4), 579–601. https://doi.org/10.1306/11020606059.

Lu, X.-C., Li, F.-C., & Watson, A. T. (1995). Adsorption measurements in Devonian Shales. *Fuel*, *74*(4), 599–603. https://doi.org/10.1016/0016-2361(95)98364-K.

Marongiu-Porcu, M., Wang, X., & Economides, M. J. (2009). Delineation of application and physical and economic optimization of fractured gas wells. In *Paper presented at the SPE production and operations symposium, Oklahoma City, Oklahoma*. https://doi.org/10.2118/120114-MS.

Maxwell, S. C., Urbancic, T. I., Steinsberger, N., & Zinno, R. (2002). Microseismic imaging of hydraulic fracture complexity in the Barnett shale. In *Paper presented at the SPE annual technical conference and exhibition, San Antonio, Texas*. https://doi.org/10.2118/77440-MS.

Mayerhofer, M. J., Lolon, E., Warpinski, N. R., Cipolla, C. L., Walser, D. W., & Rightmire, C. M. (2010). What is stimulated reservoir volume? *SPE Production & Operations*, *25*(01), 89–98. https://doi.org/10.2118/119890-PA.

Mayerhofer, M. J., Lolon, E. P., Youngblood, J. E., & Heinze, J. R. (2006). Integration of microseismic-fracture-mapping results with numerical fracture network production modeling in the Barnett shale. In *Paper presented at the SPE annual technical conference and exhibition, San Antonio, Texas, USA*. https://doi.org/10.2118/102103-MS.

McGuire, W. J., & Sikora, V. J. (1960). The effect of vertical fractures on well productivity. *Journal of Petroleum Technology*, *12*(10), 72–74. https://doi.org/10.2118/1618-G.

Medeiros, F., Ozkan, E., & Kazemi, H. (2006). A semianalytical, pressure-transient model for horizontal and multilateral wells in composite, layered, and compartmentalized reservoirs. In *Paper presented at the SPE annual technical conference and exhibition, San Antonio, Texas, USA*. https://doi.org/10.2118/102834-MS.

Meyer, B. R., Bazan, L. W., Henry Jacot, R., & Lattibeaudiere, M. G. (2010). Optimization of multiple transverse hydraulic fractures in horizontal wellbores. In *Paper presented at the SPE unconventional gas conference, Pittsburgh, Pennsylvania, USA*. https://doi.org/10.2118/131732-MS.

Mirzaei, M., & Cipolla, C. L. (2012). A workflow for modeling and simulation of hydraulic fractures in unconventional gas reservoirs. In *Paper presented at the SPE Middle East unconventional gas conference and exhibition, Abu Dhabi, UAE.* https://doi.org/10.2118/153022-MS.

Moinfar, A., Varavei, A., Sepehrnoori, K., & Johns, R. T. (2013). Development of a coupled dual continuum and discrete fracture model for the simulation of unconventional reservoirs. In *Paper presented at the SPE reservoir simulation symposium, The Woodlands, Texas, USA.* https://doi.org/10.2118/163647-MS.

Montgomery, S. L., Jarvie, D. M., Bowker, K. A., & Pollastro, R. M. (2005). Mississippian Barnett Shale, Fort Worth basin, north-central Texas: Gas-shale play with multi-trillion cubic foot potential. *American Association of Petroleum Geologists Bulletin, 89*(2), 155–175. https://doi.org/10.1306/09170404042.

Mora, C. A., & Wattenbarger, R. A. (2009). Analysis and verification of dual porosity and CBM shape factors. *Journal of Canadian Petroleum Technology, 48*(02), 17–21. https://doi.org/10.2118/09-02-17.

Morris, M. D. (1991). Factorial sampling plans for preliminary computational experiments. *Technometrics, 33*(2), 161–174. https://doi.org/10.2307/1269043.

Myers, R. H., Montgomery, D. C., & Anderson-Cook, C. M. (2016). *Response surface methodology: Process and product optimization using designed experiments.* Wiley.

Nashawi, I. S., & Malallah, A. H. (2007). Well test analysis of finite-conductivity fractured wells producing at constant bottomhole pressure. *Journal of Petroleum Science and Engineering, 57*(3), 303–320. https://doi.org/10.1016/j.petrol.2006.10.009.

Nghiem, L. X. (1983). Modeling infinite-conductivity vertical fractures with source and sink terms. *Society of Petroleum Engineers Journal, 23*(04), 633–644. https://doi.org/10.2118/10507-PA.

Nolte, K. G. (1986). Determination of proppant and fluid schedules from fracturing-pressure decline. *SPE Production Engineering, 1*(04), 255–265. https://doi.org/10.2118/13278-PA.

Nordgren, R. P. (1972). Propagation of a vertical hydraulic fracture. *Society of Petroleum Engineers Journal, 12*(04), 306–314. https://doi.org/10.2118/3009-PA.

Novlesky, A., Kumar, A., & Merkle, S. (2011). Shale gas modeling workflow: From microseismic to simulation – a horn river case study. In *Paper presented at the Canadian unconventional resources conference, Calgary, Alberta, Canada.* https://doi.org/10.2118/148710-MS.

Nuttal, B. C., Eble, C., Bustin, R. M., & Drahovzal, J. A. (2005). Analysis of Devonian black shales in kentucky for potential carbon dioxide sequestration and enhanced natural gas production. In E. S. Rubin, D. W. Keith, C. F. Gilboy, M. Wilson, T. Morris, J. Gale, et al. (Eds.), *Greenhouse gas control technologies 7* (pp. 2225–2228). Oxford: Elsevier Science Ltd.

Olorode, O., Freeman, C. M., George, M., & Blasingame, T. A. (2013). High-resolution numerical modeling of complex and irregular fracture patterns in shale-gas reservoirs and

tight gas reservoirs. *SPE Reservoir Evaluation & Engineering, 16*(04), 443–455. https://doi.org/10.2118/152482-PA.

Ozkan, E., Brown, M. L., Raghavan, R., & Kazemi, H. (2011). Comparison of fractured-horizontal-well performance in tight sand and shale reservoirs. *SPE Reservoir Evaluation & Engineering, 14*(02), 248–259. https://doi.org/10.2118/121290-PA.

Patzek, T. W., Frank, M., & Marder, M. (2013). Gas production in the Barnett Shale obeys a simple scaling theory. *Proceedings of the National Academy of Sciences, 110*(49), 19731.

Peaceman, D. W. (1978). Interpretation of well-block pressures in numerical reservoir simulation (includes associated paper 6988). *Society of Petroleum Engineers Journal, 18*(03), 183–194. https://doi.org/10.2118/6893-PA.

Peaceman, D. W. (1983). Interpretation of well-block pressures in numerical reservoir simulation with nonsquare grid blocks and anisotropic permeability. *Society of Petroleum Engineers Journal, 23*(03), 531–543. https://doi.org/10.2118/10528-PA.

Perkins, T. K., & Kern, L. R. (1961). Widths of hydraulic fractures. *Journal of Petroleum Technology, 13*(09), 937–949. https://doi.org/10.2118/89-PA.

Prats, M. (1961). Effect of vertical fractures on reservoir behavior-incompressible fluid case. *Society of Petroleum Engineers Journal, 1*(02), 105–118. https://doi.org/10.2118/1575-G.

Ross, D. J. K., & Bustin, R. M. (2007). Impact of mass balance calculations on adsorption capacities in microporous shale gas reservoirs. *Fuel, 86*(17), 2696–2706. https://doi.org/10.1016/j.fuel.2007.02.036.

Ross, D. J. K., & Bustin, R. M. (2009). The importance of shale composition and pore structure upon gas storage potential of shale gas reservoirs. *Marine and Petroleum Geology, 26*(6), 916–927. https://doi.org/10.1016/j.marpetgeo.2008.06.004.

Rubin, B. (2010). Accurate simulation of non Darcy flow in stimulated fractured shale reservoirs. In *Paper presented at the SPE western regional meeting, Anaheim, California, USA.* https://doi.org/10.2118/132093-MS.

Settari, A., Bachman, R. C., Hovem, K. A., & Paulsen, S. G. (1996). Productivity of fractured gas condensate wells - a case study of the Smorbukk field. *SPE Reservoir Engineering, 11*(04), 236–244. https://doi.org/10.2118/35604-PA.

Settari, A., Puchyr, P. J., & Bachman, R. C. (1990). Partially decoupled modeling of hydraulic fracturing processes. *SPE Production Engineering, 5*(01), 37–44. https://doi.org/10.2118/16031-PA.

Sobol, I. M. (1993). Sensitivity estimates foe nonlinear mathematical models. *MMCE, 1*(4), 8.

Storn, R., & Price, K. (1995). *Differential evolution: A simple and efficient adaptive scheme for global optimization over continuous spaces.* ICSI.

de Swaan, O.,A. (1976). Analytic solutions for determining naturally fractured reservoir properties by well testing. *Society of Petroleum Engineers Journal, 16*(03), 117–122. https://doi.org/10.2118/5346-PA.

Tannich, J. D., & Nierode, D. E. (1985). *The effect of vertical fractures on gas well productivity.* Society of Petroleum Engineers.

Tavassoli, S., Yu, W., Javadpour, F., & Sepehrnoori, K. (2013). Well screen and optimal time of refracturing: A Barnett shale well. *Journal of Petroleum Engineering, 10.* https://doi.org/10.1155/2013/817293, 2013.

Tinsley, J. M., Williams, J. R., Jr., Tiner, R. L., & Malone, W. T. (1969). Vertical fracture height-its effect on steady-state production increase. *Journal of Petroleum Technology, 21*(05), 633−638. https://doi.org/10.2118/1900-PA.

Vermylen, J. P. (2011). *Geomechanical studies of the Barnett shale, Texas, USA.* Doctor of Philosophy, Geophysics and The Committee on Graduate Studies. The Stanfoed University.

Warpinski, N., Kramm, R. C., Heinze, J. R., & Waltman, C. K. (2005). Comparison of single-and dual-array microseismic mapping techniques in the Barnett shale. In *Paper presented at the SPE annual technical conference and exhibition, Dallas, Texas.* https://doi.org/10.2118/95568-MS.

Warren, J. E., & Root, P. J. (1963). The behavior of naturally fractured reservoirs. *Society of Petroleum Engineers Journal, 3*(03), 245−255. https://doi.org/10.2118/426-PA.

Weng, X. (2015). Modeling of complex hydraulic fractures in naturally fractured formation. *Journal of Unconventional Oil and Gas Resources, 9,* 114−135. https://doi.org/10.1016/j.juogr.2014.07.001.

Weng, X., Kresse, O., Cohen, C.-E., Wu, R., & Gu, H. (2011). Modeling of hydraulic-fracture-network propagation in a naturally fractured formation. *SPE Production & Operations, 26*(04), 368−380. https://doi.org/10.2118/140253-PA.

Wu, K. (2013). Simultaneous multi-frac treatments:fully coupled fluid flow and fracture mechanics for horizontal wells. In *Paper presented at the SPE annual technical conference and exhibition, New Orleans, Louisiana, USA.* https://doi.org/10.2118/167626-STU.

Wu, K. (2014). *Numerical modeling of complex hydraulic fracture development in unconventional reservoirs.* Doctor of Philosophy. The University of Texas at Austin.

Wu, R., Kresse, O., Weng, X., Cohen, C.-E., & Gu, H. (2012). Modeling of interaction of hydraulic fractures in complex fracture networks. In *Paper presented at the SPE hydraulic fracturing technology conference, The Woodlands, Texas, USA.* https://doi.org/10.2118/152052-MS.

Wu, Kan, & Olson, J. E. (2013). Investigation of the impact of fracture spacing and fluid properties for interfering simultaneously or sequentially generated hydraulic fractures. *SPE Production & Operations, 28*(04), 427−436. https://doi.org/10.2118/163821-PA.

Xu, W., Thiercelin, M. J., Ganguly, U., Weng, X., Gu, H., Onda, H., et al. (2010). Wiremesh: A novel shale fracturing simulator. In *Paper presented at the international oil and gas conference and exhibition in China, Beijing, China.* https://doi.org/10.2118/132218-MS.

Xu, G., & Wong, S.-W. (2013). Interaction of multiple non-planar hydraulic fractures in horizontal wells. In *Paper presented at the international petroleum technology conference, Beijing, China.* https://doi.org/10.2523/IPTC-17043-MS.

Yang, C., Card, C., & Nghiem, L. (2009). Economic optimization and uncertainty assessment of commercial SAGD operations. *Journal of Canadian Petroleum Technology, 10*(09), 33−40. https://doi.org/10.2118/09-09-33.

Yang, C., Nghiem, L. X., Card, C., & Bremeier, M. (2007). Reservoir model uncertainty quantification through computer-assisted history matching. In *Paper presented at the SPE annual technical conference and exhibition, Anaheim, California, USA.* https://doi.org/10.2118/109825-MS.

Yew, C. H., & Weng, X. (2014). *Mechanics of hydraulic fracturing.* Elsevier Science.

Yousefzadeh, A., Qi, L., Virues, C., & Aguilera, R. (2017). Comparison of PKN, KGD, Pseudo3D, and diffusivity models for hydraulic fracturing of the horn river basin shale gas formations using microseismic data. In *Paper presented at the SPE unconventional resources conference, Calgary, Alberta, Canada.* https://doi.org/10.2118/185057-MS.

Yu, W. (2015). *Developments in modeling and optimization of production in unconventional oil and gas reservoirs.* Doctor of Philosophy. The University of Texas at Austin.

Yu, W., Gao, B., & Sepehrnoori, K. (2014). Numerical study of the impact of complex fracture patterns on well performance in shale gas reservoirs. *Journal of Petroleum Science Research, 3*(2). https://doi.org/10.14355/jpsr.2014.0302.05.

Yu, W., & Sepehrnoori, K. (2013). Optimization of multiple hydraulically fractured horizontal wells in unconventional gas reservoirs. In *Paper presented at the SPE production and operations symposium, Oklahoma City, Oklahoma, USA.* https://doi.org/10.2118/164509-MS.

Yu, W., & Sepehrnoori, K. (2014). An efficient reservoir-simulation approach to design and optimize unconventional gas production. *Journal of Canadian Petroleum Technology, 53*(02), 109−121. https://doi.org/10.2118/165343-PA.

Yu, W., Sepehrnoori, K., & Patzek, T. W. (2016). Modeling gas adsorption in Marcellus shale with Langmuir and BET isotherms. *SPE Journal, 21*(02), 589−600. https://doi.org/10.2118/170801-PA.

Yu, W., Varavei, A., & Sepehrnoori, K. (2015). Optimization of shale gas production using design of experiment and response surface methodology. *Energy Sources, Part A: Recovery, Utilization, and Environmental Effects, 37*(8), 906−918. https://doi.org/10.1080/15567036.2013.812698.

Zhang, X., Du, C., Deimbacher, F., Martin, C., & Harikesavanallur, A. (2009). Sensitivity studies of horizontal wells with hydraulic fractures in shale gas reservoirs. In *Paper presented at the international petroleum technology conference, Doha, Qatar.* https://doi.org/10.2523/IPTC-13338-MS.

Zheltov, A. K. (1955). 3. Formation of vertical fractures by means of highly viscous liquid. In *Paper presented at the 4th world petroleum congress, Rome, Italy.*

Zhou, W., Banerjee, R., Poe, B. D., Spath, J., & Thambynayagam, M. (2013). Semianalytical production simulation of complex hydraulic-fracture networks. *SPE Journal, 19*(01), 6−18. https://doi.org/10.2118/157367-PA.

Zimmerman, R. W., Chen, G., Hadgu, T., & Bodvarsson, G. S. (1993). A numerical dual-porosity model with semianalytical treatment of fracture/matrix flow. *Water Resources Research, 29*(7), 2127−2137. https://doi.org/10.1029/93WR00749.

CHAPTER 5

Challenges of Shale Reservoir Technologies

ABSTRACT

Recently, CO_2 injection technique has attracted attention as a potential method to improve oil and gas recovery and to store greenhouse gas in shale reservoirs. For successful application of CO_2 injection methods in shale reservoirs, transport mechanisms should be understood comprehensively. Based on shale reservoir models generated in Chapter 4, CO_2 injection models considering non-Darcy flow, multilayer adsorption, molecular diffusion, stress-dependent deformation, and phase behavior change in nanopores are presented. In addition, for more accurate simulation of shale reservoirs, organic matter pores and inorganic matter pores should be distinguished. Organic and inorganic pores have different flow mechanisms because of distinctive pore structure, adsorption capacity, and geomechanic properties. Several studies considering effects of organic matters in numerical modeling of shale reservoirs are presented.

MULTICOMPONENT TRANSPORT IN CO_2 INJECTION PROCESS

In spite of the sizable growth of shale resource production, oil and gas recovery in respect of single shale well is still low compared with that of a single well in conventional reservoirs. Although most shale gas and oil wells present a steep increase of production rate in the early stage, a rapid decline of production is observed just after a few months. Fig. 5.1 presents type gas well decline curves for Haynesville, Marcellus, Barnett, Fayetteville, and Woodford (Hughes, 2013). As shown in Fig. 5.1, average gas production rate decreases 84% during the first 3 years after production started. Fig. 5.2 shows Bakken and Three Forks type oil well decline curves in North Dakota (Hughes, 2013). In the Bakken and Three Forks oil reservoirs, the production rates decrease 71% and 70% in the first year, and the 3-year declines of production rates are 86% and 85%, respectively. In shale reservoirs, gas and oil recovery commonly remains only from 15% to 25% and from 3% to 10%, respectively. Therefore,

in recent years, CO_2 injection has attracted significant attention as a potential method to enhance hydrocarbon recovery in shale reservoirs. Several field injection tests with water and CO_2 were conducted in Bakken reservoir (Sorensen & Hamling, 2016). In these tests, water and gas injectivity into various lithofacies has been demonstrated. However, there was no obvious incremental improvement in oil production with water injection. In CO_2 injection cases, production responses to injection were observed although those responses were not related to higher oil production. These results suggested that fluid flow in shale reservoir can be influenced by CO_2 injection so that enhanced gas recovery (EGR) and enhanced oil recovery (EOR) in shale reservoirs are possible (Sorensen & Hamling, 2016). Development of effective EGR and EOR approaches coupled with CO_2 injection will require detailed study of transport mechanisms involved in shale reservoirs as well as reservoir and fracture conditions.

Several reports have shown that the affinity of CO_2 adsorption to the shale reservoir is higher than that of CH_4 under subsurface conditions depending on the thermal maturity of the organic materials (Busch et al., 2008; Shi & Durucan, 2008; Vermylen, 2011). The higher affinity of CO_2 to the shale reservoir could initiate mechanisms to desorb the CH_4 originally existed and to adsorb and trap the CO_2 suspected of contributing to the greenhouse effect as shown in Fig. 5.3 (Godec, Koperna, Petrusak, & Oudinot, 2014). Therefore, CO_2 injection in shale reservoirs contributes not only to the additional gas and oil production but also to the geological sequestration of CO_2, which otherwise causes a greenhouse effect. Although CO_2 injection is common in conventional reservoirs in these days, it is expected to behave differently in shale reservoirs so that accurate understanding of fluid flow and transport during CO_2 injection is required.

There have been numerous researches that analyze the effects of CO_2 injection in shale gas reservoirs. Feasibility studies of CO_2 injection in shale gas reservoirs were performed by Schepers, Nuttall, Oudinot, and

Transport in Shale Reservoirs. https://doi.org/10.1016/B978-0-12-817860-7.00005-X

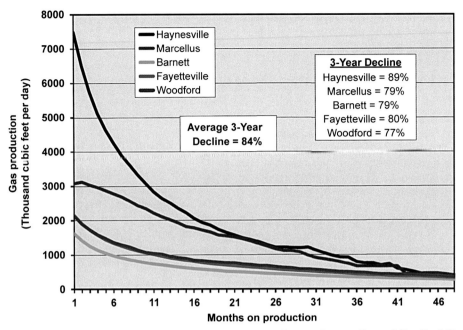

FIG. 5.1 Type gas well decline curves for Haynesville, Marcellus, Barnett, Fayetteville, and Woodford. (Credit: Hughes, J. D. (2013). *Tight oil: A solution to U.S. import dependence?*. Denver, Colorado, USA: Geological Society of America, October 28.)

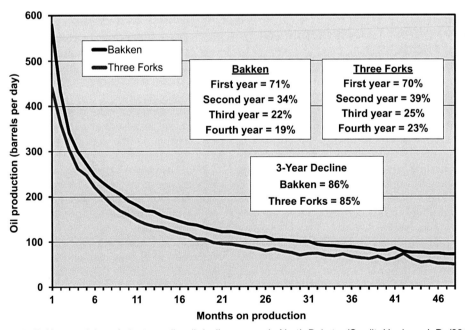

FIG. 5.2 Bakken and three forks type oil well decline curves in North Dakota. (Credit: Hughes, J. D. (2013). *Tight oil: A solution to U.S. import dependence?* Denver, Colorado, USA: Geological Society of America, October 28.)

Stage 1   Stage 2   Stage 3

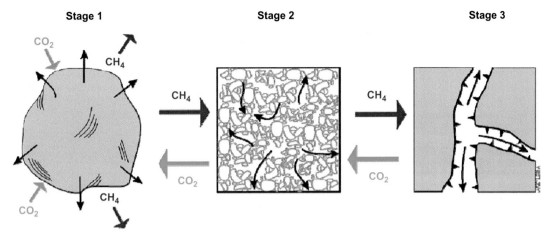

FIG. 5.3 Schematic view of the flow dynamics of CO_2 and CH_4 in shale gas reservoirs. (Credit: Godec, M., Koperna, G., Petrusak, R., & Oudinot, A. (2014). Enhanced gas recovery and CO_2 storage in gas shales: A summary review of its status and potential. *Energy Procedia*, 63, 5849–5857. doi:10.1016/j.egypro.2014.11.618.)

Gonzalez (2009) and Kalantari-Dahaghi (2010). Godec, Koperna, Petrusak, and Oudinot (2013; 2014) performed the reservoir simulation of CO_2 injection in Marcellus shale. Simulation results from their studies indicated that gas production could be increased by 7% using CO_2 injection at the optimal spacing between injection and production wells. Liu, Ellett, Xiao, John, and Rupp (2013) focused on CO_2 storage in Devonian and Mississippian New Albany shale gas plays. They showed that more than 95% of injected CO_2 was sequestered instantaneously with gas adsorption being the dominant storage mechanism. These studies presented the potential of CO_2 injection with simple shale models, which do not consider any complex transport mechanisms. Fathi and Akkutlu (2014) presented simulations of multicomponent transport between CO_2 and CH_4 in shale reservoirs. Their approach included competitive transport and adsorption effects during CO_2 injection, but they only considered the huff and puff scenario. Eshkalak, Al-shalabi, Sanaei, Aybar, and Sepehrnoori (2014) examined both CO_2 huff and puff and CO_2 flooding and insisted that the huff and puff process is not a viable option for CO_2 EGR. However, according to extensive numerical simulations presented in the following section, huff and puff method also could be a useful option for CO_2 EGR, depending on fracture conductivity and well spacing. Jiang, Shao, and Younis (2014) developed a fully coupled multicontinuum multicomponent simulator considering Langmuir adsorption and deformation of shale reservoir for the CO_2 EGR and CO_2 storage process. However, they applied only pressure-dependent permeability of

fractures without detailed geomechanic effects in synthetic shale reservoir model.

There is also no consensus as to whether CO_2 injection in shale oil reservoirs is feasible or not because research in this reservoir is in its early stage. Several researchers have tried to investigate how CO_2 injection can be performed in lab-scale experiments of shale cores. Kovscek, Robert, Tang, and Vega (2008) and Vega, O'Brien, and Kovscek (2010) conducted experiments to investigate the effects of CO_2 injections in siliceous shale. They showed that the improvement of oil recovery is sensitive to the distribution of the gas phase near the fracture face. They indicated that both diffusion and convective-dispersion mechanisms were significant in the experimental oil displacement. Vega et al. (2010) constructed a simulation model with experimental conditions, but they could not obtain an exact match of the experimental results. Through CO_2 huff and puff experiments with a small piece of Bakken shale rock, Hawthorne et al. (2013) showed nearly complete oil recovery from the Middle Bakken reservoir in spite of ultralow permeability. Gamadi, Sheng, and Soliman (2013; 2014) and Tovar, Eide, Graue, and Schechter (2014) performed a similar experimental study on repeated huff and puff gas injection in shale cores. They found that oil recovery increased by approximately 10%–50%, depending on the experimental conditions. Numerous simulation studies of CO_2 injection in tight and shale reservoirs have shown that CO_2 injection is effective for improving oil recovery (Chen, Balhoff, & Mohanty, 2014; Pu & Li, 2015; Sheng, 2015a, 2015b; Yu, Lashgari, Wu, & Sepehrnoori,

2015). Ghorbae and Alkhansa (2012) studied diffusive exchanges of components between the matrix and fracture, which significantly contributed to oil recovery from the matrix in the fractured reservoirs. To evaluate the importance of molecular diffusion, they simulated laboratory experiments using a compositional model. They concluded that neglecting molecular diffusion resulted in the underestimation of the gas injection efficiency. In a shale reservoir, the effects of molecular diffusion would be more significant and should be considered to simulate the CO_2 EOR process accurately. Alharthy et al. (2015) presented laboratory and reservoir modeling of CO_2 EOR with Bakken data. A numerical model was constructed to match laboratory oil recovery results, and it was scaled up to field applications. They considered only the CO_2 huff and puff method and did not include complex transport mechanisms except for molecular diffusion. Kalra, Tian, and Wu (2018) and Wan and Liu (2018) presented the simulation of CO_2 injection considering only pressure-dependent permeability for EOR in shale oil reservoirs. Using a single porosity model, Zhang, Yu, Li, and Sepehrnoori (2018) studied the effects of diffusion, nanopore confinement, and pressure-dependent deformation on the effectiveness of CO_2 EOR in shale oil reservoir. Because the confinement effect is observed at nanopores of shale reservoirs, the suggested single porosity shale oil model seems to be inadequate for considering nanopore confinement effect. In addition, the dual porosity model should be considered because of the significant impacts of natural fractures in shale reservoirs.

Especially in recent years, interests to CO_2 injection in shale reservoir have been explosively grown, whereas integrative understanding of CO_2 injection process considering complex transport mechanisms is still deficient. Combined mechanisms of stress-dependent compaction coupled with adsorption isotherm, multicomponent and multilayer adsorption, nanopore confinement, molecular diffusion, and complex flow through hydraulic fracture system should be considered comprehensively. In the following section, integrative shale reservoir models are presented to evaluate CO_2 injection. CO_2 flooding and huff and puff methods were used to assess the possibility of hydrocarbon improvement and CO_2 storage in shale reservoirs. In Barnett shale gas reservoir and Bakken shale oil reservoir models, effects of stress-dependent compaction, molecular diffusion, and competitive adsorption were investigated during CO_2 injection processes. In the last section, to consider realistic horizontal wells and hydraulic fracture system in the shale reservoirs, complex hydraulic fracture model generated by fracturing simulator GOHFER (Barree & Associates, 2015) was used. In the complex hydraulic fracture model, where CO_2 flooding was performed, effects of fracture properties for oil recovery were investigated. A series of reservoir simulations with the developed model will suggest systematic and comprehensive understanding and evaluation for CO_2 injection in shale reservoirs.

Shale Gas Reservoirs

To analyze the realistic effects of CO_2 injection in a shale gas reservoir, a numerical model of Barnett shale reservoir was used. Barnett shale reservoir model was generated by the history-matching technique as shown in Chapter 4.2. Using the matched values, a segment of Barnett shale reservoir ($330 \times 510 \times 330$ ft^3), which was simplified for computational efficiency, was generated including two hydraulically fractured horizontal wells (Fig. 5.4). In a multifractured horizontal well,

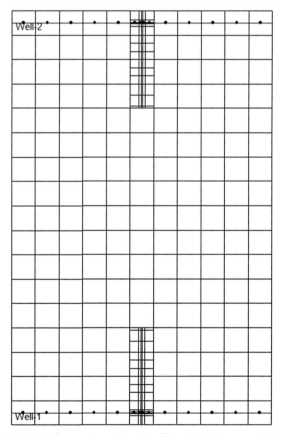

FIG. 5.4 Segment model of the Barnett shale reservoir to simulate the CO_2 injection processes.

hydraulic fractures are repeated through the well so that simplification performed in this study does not affect the results with the assumptions that reservoir and natural fractures are homogeneous and properties of hydraulic fractures are same. The reservoir assumed to be isothermal and no flow boundary conditions to consider one of the repeated segments. Initial water saturation is 0.2, but water is immobile in the matrix. As shown in Fig. 5.5, CO_2 adsorption data fit better with BET isotherm than Langmuir isotherm, especially in high pressure. Based on fitting results, BET isotherm is applied in this model. To analyze the effects of the geomechanic model, power law correlation was considered.

In the Barnett shale model, three cases were compared to analyze the effects of CO_2 injection in shale reservoirs, such as CO_2 flooding, huff and puff, and no injection scenarios. In the model without CO_2 injection, both wells are produced continuously over 30 years. In CO_2 flooding model, two horizontal wells are produced for the first 5 years. Then, CO_2 is injected in one well with the constant rate of 100 Mscf/day, whereas the other well continues to produce. The CO_2 injection rate is set to consider the efficiency of both CH_4 production and CO_2 storage. During CO_2 injection, pressure near injection well increases up to initial reservoir pressure. Because of reservoir condition, injected CO_2 is a supercritical state. After 5 years of injection, the CO_2 injector is shut in, and the other well is produced for an additional 20 years. In the huff and puff model, CO_2 is injected to both wells with a rate of 250 Mscf/day for 1 month after the first 5 years of primary production. After another 1 month of soaking period, both wells are produced for 4 months. This cycle is repeated for 6 years to inject the same amount of CO_2 with flooding case.

To analyze the effects of CO_2 injection in Barnett shale reservoir, CH_4 production of Barnett shale model was presented with three different scenarios (Fig. 5.6). The CH_4 production with CO_2 flooding is lower than that of models without CO_2 injection in the early stage of CO_2 injection. This is a transitory phenomenon due to the distance between injection and production wells. After the production well is switched to the injection well, CO_2 flows into the production well slowly so that effects of CO_2 on CH_4 production are also turned up slowly. Increased CH_4 production by CO_2 injection is observed approximately 1 year after the injection started. After this early period, the CH_4 production of the model with CO_2 flooding increases rapidly and overtakes the CH_4 production of the model without CO_2 injection. The process of CO_2 EGR is mainly dominated by pressurizing effect and CH_4 desorption induced by CO_2 adsorption. The pressurizing effect caused by CO_2 injection also induces rock deformation so that permeability increases. The CH_4 production of the model with CO_2 huff and puff increases steadily. Because the CO_2 cannot spread to the far reservoir and it is produced immediately after injection, CH_4 production of CO_2 huff and puff case is lower than that of CO_2 flooding case. Fig. 5.6 shows that the CH_4 production of the models with CO_2 flooding and huff and puff

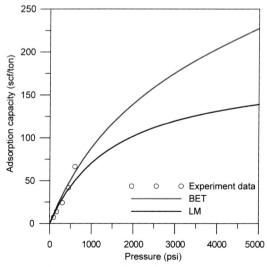

FIG. 5.5 Fitting curves of CO_2 adsorption experiment data with Langmuir and BET isotherms.

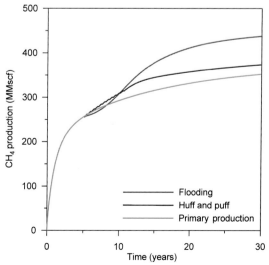

FIG. 5.6 CH_4 production for the Barnett shale models with CO_2 flooding, huff and puff, and no CO_2 injection scenarios.

is 24% and 6% higher than that of the model without CO_2 injection at the end of the production, respectively.

Fig. 5.7 shows the CO_2 production and storage of the Barnett shale models with CO_2 flooding and huff and puff scenarios. Amount of injected CO_2 is the same in both cases. In CO_2 flooding scenario, because only 1% of the injected CO_2 is produced, 99% of injected CO_2 is stored in the reservoir at the end of the production. Although CO_2 breakthrough is observed 3 years after the injection started, the production rate of CO_2

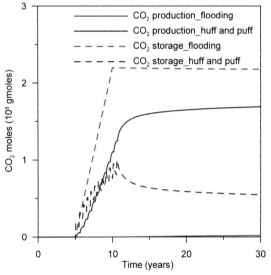

FIG. 5.7 CO_2 production and storage of the Barnett shale models with CO_2 flooding and huff and puff scenarios.

remains extremely low. In CO_2 huff and puff scenario, however, the 75% of injected CO_2 is produced, and only 25% of injected CO_2 is stored. Owing to cyclic injection and production in the same well, CO_2 cannot spread to the far reservoir so that significant portion of the injected CO_2 is produced compared with CO_2 flooding case as shown in Fig. 5.8. Figs. 5.9 and 5.10 provide the classifications of the stored CO_2 such as free, adsorbed, and dissolved CO_2 for the Barnett shale reservoir with CO_2 flooding and huff and puff scenarios. In the case of CO_2 flooding scenario shown in Fig. 5.9, the amount of free CO_2 is highest directly after the end of CO_2 injection. As time elapses, the free CO_2 is displaced by pressure differences and molecular diffusion effect. This free CO_2 is partially adsorbed on the surface or dissolved in the water so that the amount of free CO_2 decreases. Eventually, the injected CO_2 is stored as free, adsorbed, and dissolved states in proportions of 42%, 55%, and 3%, respectively, at the end of the production. Among these states of CO_2, the free CO_2 is mobile. The adsorbed and dissolved CO_2 are immobile because they are trapped in the surface of rock particles and connate water, respectively. The trapped CO_2 is important for storage because of its stability for long-term sequestration. In the case of CO_2 flooding scenario, 58% of stored CO_2 is permanently trapped. Fig. 5.10 shows the classifications of stored CO_2 in the case of CO_2 huff and puff scenario. Because CO_2 cannot spread to the far reservoir, all states of CO_2 decrease after huff and puff process is finished. At the end of the production, CO_2 is stored as free, adsorbed, and dissolved state in proportions of 39%, 58%, and 3%, respectively.

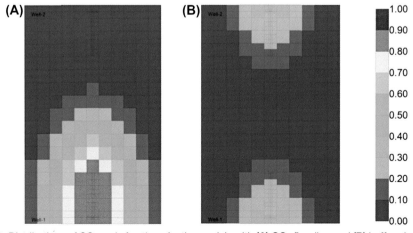

FIG. 5.8 Distributions of CO_2 mole fractions for the models with **(A)** CO_2 flooding and **(B)** huff and puff at the after 5 years of CO_2 injection.

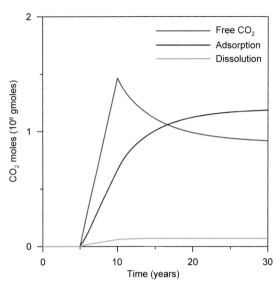

FIG. 5.9 Free, adsorbed, and dissolved CO_2 in the Barnett shale model with CO_2 flooding scenario.

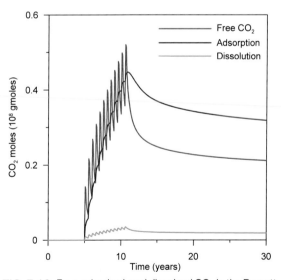

FIG. 5.10 Free, adsorbed, and dissolved CO_2 in the Barnett shale model with CO_2 huff and puff scenario.

Fig. 5.11 provides the adsorption of CH_4 and CO_2 for Barnett shale models with CO_2 flooding, huff and puff, and no injection scenarios. Red, blue, and green lines present the model with CO_2 flooding, huff and puff, and no CO_2 injection scenarios, respectively. Solid and dashed lines indicate the amounts of CH_4 and CO_2 adsorption, respectively. In the model

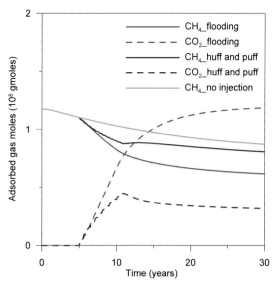

FIG. 5.11 Adsorption of CH_4 and CO_2 for the Barnett shale models with CO_2 flooding, huff and puff, and no injection scenarios.

without CO_2 injection, only 26% of initially adsorbed CH_4 is desorbed during production. On the contrary, in the models with CO_2 flooding and huff and puff scenarios, desorption of CH_4 is activated by competitive sorption with CO_2. As shown in Fig. 5.11, 48% and 32% of initially adsorbed CH_4 are desorbed by CO_2 flooding and huff and puff techniques. Fig. 5.12 shows distributions of the CH_4 adsorption for the Barnett shale model with these three scenarios. Figs. 5.12A–5.12C indicate CH_4 adsorption after 10 years, and Figs. 5.12D–5.12F show CH_4 adsorption at the end of the simulation. In case of CO_2 flooding model, a significant amount of CH_4 is desorbed near the injection well in early stage, and CH_4 is desorbed in a larger area as CO_2 flows to the production well as time goes on. In the model of CO_2 huff and puff, however, desorption of CH_4 is limited to the nearby region of each hydraulic fracture so that amount of CH_4 desorption is low. In other words, the amounts of CO_2 adsorption and CH_4 desorption are large when CO_2 flooding is performed in the Barnett shale reservoir.

Fig. 5.13 presents CO_2 mole fractions for the models with and without consideration of molecular diffusion at the end of the simulation. Fig. 5.13A, which shows the model considering molecular diffusion, presents more widely spread CO_2 in the reservoir than the model not considering molecular diffusion, shown in Fig. 5.13B. The results indicate that molecular diffusion

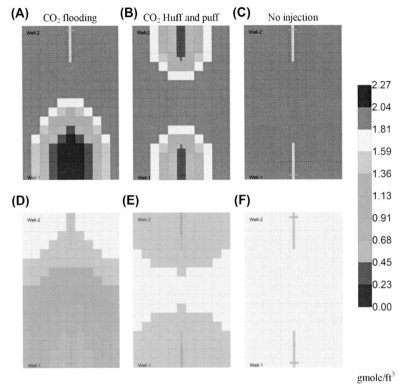

FIG. 5.12 Distributions of the adsorbed CH_4 at **(A–C)** 10 and **(D–F)** 30 years for **(A, D)** CO_2 flooding, **(B, E)** huff and puff, and **(C, F)** no injection scenarios.

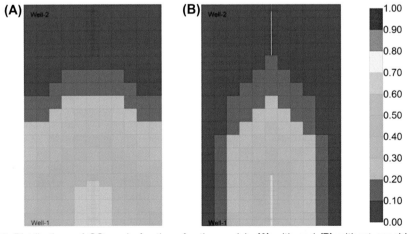

FIG. 5.13 Distributions of CO_2 mole fractions for the models **(A)** with and **(B)** without consideration of molecular diffusion at the end of the simulation.

enables CO_2 to displace CH_4 smoothly under ultralow matrix permeability conditions. Because of the ultralow permeability of the shale reservoir, the effect of diffusion is much higher than conventional reservoirs.

Therefore, it should be considered in the CO_2 injection model of the shale reservoir.

The conductivity of a fracture network in a shale reservoir is sensitive to the changes in stress and strain

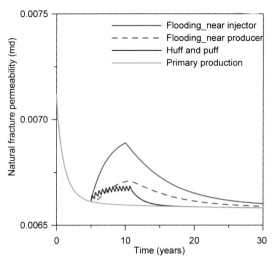

FIG. 5.14 Natural fracture permeability of the Barnett shale models with CO_2 flooding, huff and puff, and no injection scenarios.

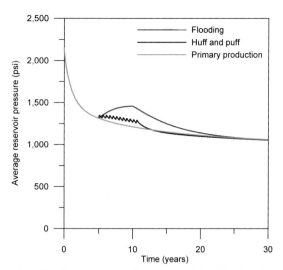

FIG. 5.15 Average reservoir pressure of the CO_2 flooding, huff and puff, and primary production cases in the Barnett models.

during production. Therefore, geomechanic effects during production must be included to simulate the behavior of a shale gas reservoir accurately (Cho, Ozkan, & Apaydin, 2013). To calculate the geomechanic effects, stress-dependent permeability coupled with geomechanics is applied in this model. Power law correlations are used to compute the stress-dependent properties. Fig. 5.14 shows the changes in natural fracture permeability of adjacent wells in the Barnett shale models with CO_2 flooding, huff and puff, and no injection scenarios. In the first 5 years, permeability decreases rapidly due to the decrease of pressure during production. After CO_2 injection begins, natural fracture permeability increases until the injection is stopped and then decreases again. The increments of permeability described by the geomechanic model have a positive effect on the CO_2 injection in the shale reservoir. In the case of CO_2 flooding, because of pressurizing effect of CO_2 injection, natural fracture permeability of near injector is higher than that of near producer. The increase of natural fracture permeability in CO_2 huff and puff scenario is lower than that in CO_2 flooding scenario even near producer. The pressurizing effect of CO_2 huff and puff case is low so that permeability could not increase as CO_2 flooding case. Fig. 5.15 presents average reservoir pressure of each model, and it corresponds with permeability change. The geomechanic effect is positive for CO_2 injection process, and especially CO_2 flooding scenario shows considerable permeability increase in Barnett shale model.

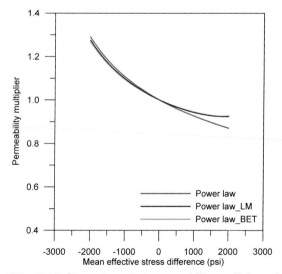

FIG. 5.16 Stress-dependent permeability multiplier of models with and without Langmuir and BET adsorption coupling.

Fig. 5.16 presents permeability multiplier depending on effective stress difference. Results from models with only power law correlation, power law correlation coupled with Langmuir adsorption, and power law correlation coupled with BET adsorption are presented. As shown in Fig. 5.17, in the models with deformation and adsorption coupling, permeability increases due to desorption, and permeability decreases due to adsorption. However, permeability change by adsorption

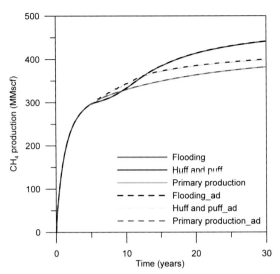

TABLE 5.1
Parameters for Sensitivity Analysis of CO$_2$ Injection in Barnett Shale Gas Reservoir

Uncertain Parameters	Minimum	Maximum
Matrix porosity	0.04	0.07
Matrix permeability (md)	1×10^{-7}	1×10^{-5}
Natural fracture permeability (md)	0.001	0.01
Hydraulic fracture permeability (md)	10,000	100,000
Hydraulic fracture half-length (ft)	30	210
Young's modulus (psi)	3,000,000	6,000,000
Poisson's ratio	0.2	0.3

FIG. 5.17 CH$_4$ production for the Barnett shale model with and without adsorption coupling in stress-dependent permeability correlation.

effects is small compared with stress effects, especially in the general range of effective stress variation. Fig. 5.17 shows CH$_4$ production depending on considering BET adsorption coupling in stress-dependent correlation. Although permeability changes due to adsorption coupling, effects of adsorption on permeability is insignificant compared with stress effects. Consequently, CH$_4$ production difference between models with and without effects of adsorption on permeability is less than 1%.

Shale gas reservoirs show high uncertainty because of the inestimable reservoir and fracture properties. Owing to these uncertainties, effects of reservoir and fracture properties should be rigorously investigated for the field application of CO$_2$ injection process. Sensitivity analysis was performed for CH$_4$ production and CO$_2$ storage. Table 5.1 presents uncertainty parameters used for sensitivity analysis. Fig. 5.18 provides results of sensitivity analysis for the objective functions of CH$_4$ production and CO$_2$ storage. In this study, matrix porosity, matrix permeability, natural fracture permeability, hydraulic fracture permeability, hydraulic fracture half-length, Young's modulus, and Poisson's ratio are considered as parameters for sensitivity analysis. Fig. 5.18 shows the influence of each parameter on CH$_4$ production and CO$_2$ storage. For the CO$_2$-EGR, natural fracture permeability, matrix porosity, and hydraulic fracture half-length are of importance. The conductivity of the fracture system between producer and injector shows positive influence on EGR so that

investigation of the fracture system should be preceded for field application of CO$_2$ injection in shale reservoir. Young's modulus affects the deformation of shale rock. As Young's modulus increases, CH$_4$ production decreases owing to high compaction of shale rock. For the CO$_2$ storage, influences of uncertain parameters are small compared with CO$_2$-EGR. Hydraulic fracture half-length, natural fracture permeability, and hydraulic fracture permeability are relatively important. Especially, hydraulic fracture half-length presents adverse effects on CO$_2$ storage while it shows a positive impact on CH$_4$ production. Therefore, the influence of fracture system should be considered depending on the objectives of CO$_2$ injection in shale gas reservoirs.

Shale Oil Reservoir

Based on the CO$_2$ huff and puff experiment and simulation performed by Alharthy et al. (2015), Bakken core model was constructed to investigate the effects of CO$_2$ injection in shale rocks. In the experiment, a middle Bakken core was placed in the center of a cylindrical extraction vessel. Fractures surrounding the matrix are described by the space between the core and the extraction vessel wall. CO$_2$ was injected at 5000 psi into the inlet valve, and pressure was maintained constantly during the entire experiment. The outlet valve was closed for 50 min to soak the core with injected CO$_2$, and then the outlet valve was opened to produce oil for 10 min. This cycle was repeated for approximately 500 min. Based on this experiment, a numerical core model was constructed. Fig. 5.19 shows schematic views of the core model with radial grids. The red cells are the space in the extraction vessel indicating fractures and

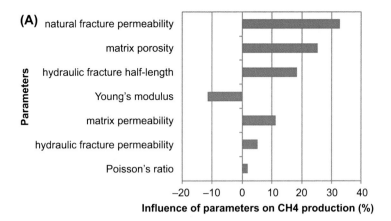

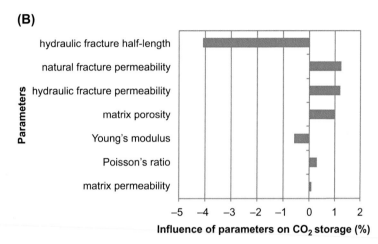

FIG. 5.18 Result of sensitivity analysis for **(A)** enhanced CH4 recovery and **(B)** CO2 storage.

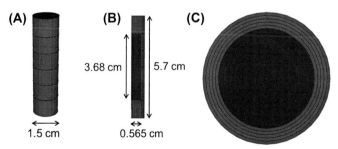

FIG. 5.19 Schematic views of the numerical Bakken core model with radial grids.

the blue cells indicate the Bakken core. Porosity and permeability of core are 0.08 and 0.043 md, respectively. The core has dimensions of length 3.68 cm and diameter 1.13 cm. Dimensions of extraction vessel length and diameter are 5.7 and 1.5 cm.

Using this core model, history matching was performed. To investigate the effects of molecular diffusion in the Bakken formation, three different diffusion cases were considered to match oil recovery data (Fig. 5.20). The matched results with Wilke-Chang and Sigmund diffusion correlations and without molecular diffusion were presented with experimental oil recovery data. In these results, the model without the molecular diffusion mechanism shows the largest matching error,

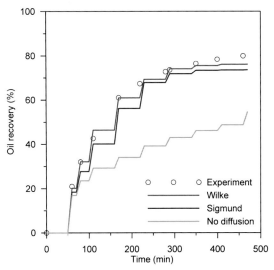

FIG. 5.20 Oil recovery data and the history matching results of the Bakken core with Wilke-Chang, Sigmund, and no diffusion models.

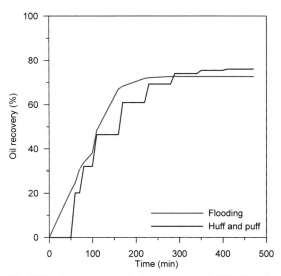

FIG. 5.21 Comparison of oil recovery with CO_2 flooding and huff and puff cases in the Bakken core model.

whereas models with the molecular diffusion present smaller errors. These results illustrate that molecular diffusion should be considered for exact history matching. Owing to the low permeability, CO_2 transport is affected considerably by molecular diffusion in shale oil reservoirs. In the Bakken core, the Wilke-Chang correlation presents more accurate matching than the Sigmund correlation. Therefore, the Wilke-Chang correlation was applied to simulate the molecular diffusion mechanism in the base Bakken reservoir model. In the core model, CO_2 flooding with a constant injection rate was tested and compared with the CO_2 huff and puff process (Fig. 5.21). Using the same amount of injected CO_2, similar oil recovery was achieved with both methods. The result shows the potential of both CO_2 flooding and the huff and puff methods to improve oil production in shale reservoirs.

To analyze the effects of CO_2 injection in shale oil reservoirs, a numerical model was constructed. The model was based on field data from the Bakken Formation, published by Yu et al. (2015). To reduce computational cost, the segment of the Bakken reservoir was constructed with two horizontal wells and hydraulic fractures within one stage (Fig. 5.22). The size of the segment is set to $340 \times 900 \times 40 \ ft^3$, with 80 and 900 ft of hydraulic fracture spacing and well spacing, respectively. In this model, the matrix and natural fracture system of the Bakken was generated by a dual porosity/dual permeability model. To reduce numerical dispersion, a local grid refinement technique was used

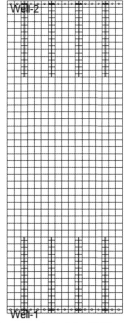

FIG. 5.22 Schematic view of the Bakken reservoir model.

to set up the grid of hydraulic fractures. Reservoir and fracture properties of the Bakken Formation are presented in Table 5.2. Typical fluid properties of the Bakken Field, presented by Yu et al. (2015), are used in this model. The oil gravity is roughly $42°$API, which denotes light oil. Seven different pseudocomponents of

TABLE 5.2
Properties of Reservoir and Fracture Used in the Bakken Shale Reservoir Model

Reservoir Pressure (psi)	8,000
Reservoir temperature (°F)	240
Matrix porosity	0.07
Matrix permeability (md)	0.01
Total compressibility (psi^{-1})	1×10^{-6}
Hydraulic fracture permeability (md)	50,000
Hydraulic fracture width (ft)	0.001
Hydraulic fracture half-length (ft)	210
Hydraulic fracture height (ft)	40

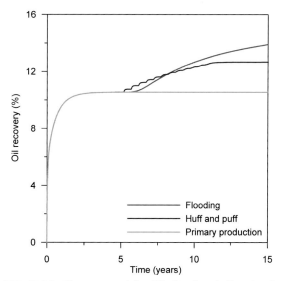

FIG. 5.23 Oil recovery of the CO_2 flooding, huff and puff, and no injection cases in the Bakken models.

Bakken crude oil include CO_2, N_2, CH_4, C_2–C_4, C_5–C_7, C_8–C_9, and C_{10+}; specific fluid properties are also provided by Yu et al. (2015). Based on the results of core test, the Wilke-Chang correlation was used to consider molecular diffusion in the model. Geomechanic effects were considered by using a linear elastic model coupled with a stress-dependent exponential correlation. The experimental coefficients were obtained from Cho et al. (2013). Stress and strain are calculated by coupling of mass conservation, and displacement equations and permeability are computed by multipliers depending on the stress variation.

In the Bakken model, CO_2 flooding and huff and puff processes were applied to compare oil production and CO_2 storage. During the primary production period in the first 5 years, both horizontal wells were produced. After the primary production period, in the model with CO_2 flooding, CO_2 was injected into one well while the other well continued to produce oil and gas. After 1 year of CO_2 injection, the injection well was shut in, and the other well produced for an additional 9 years. In the CO_2 huff and puff process, CO_2 was injected into both wells after primary production. After 1 month of CO_2 injection, both wells were shut in for 1 month to allow soaking, and then the wells produced for 4 months. The procedure, which is one cycle of the CO_2 huff and puff process, was repeated for 6 years. In both methods, the amount of total injected CO_2 is equal.

In the proposed Bakken reservoir model, the effects of CO_2 injection on oil recovery improvement and CO_2 storage were analyzed. Fig. 5.23 shows the oil recovery of the model with the CO_2 flooding, huff and puff, and no injection cases. In the early stage of CO_2

injection, the oil recovery from CO_2 huff and puff is higher than from CO_2 flooding. In the huff and puff model, an improvement in oil recovery is observed immediately after the first cycle. For the flooding model, the improvement of oil recovery is observed slowly owing to the unstimulated zone, which is not affected by hydraulic fracturing between the injector and producer. However, after 3 years of CO_2 injection, oil recovery of CO_2 flooding case exceeds that of CO_2 huff and puff case. At the end of the simulation, oil recoveries of models with the CO_2 flooding and huff and puff methods are 13.9% and 12.6%, respectively. Compared with primary oil production, oil recovery increases 31.7% and 20.0% in the cases of CO_2 flooding and huff and puff, respectively. According to Fig. 5.24, the amounts of CO_2 production and storage in the CO_2 flooding and huff and puff cases are significantly different in spite of the same amount of CO_2 being injected. At the end of the simulation, only a small amount of injected CO_2 is produced in the CO_2 flooding model, whereas 37% of the injected CO_2 is produced in the CO_2 huff and puff model. Consequently, most of injected CO_2 is stored with the flooding method, while only 63% of injected CO_2 is stored with the huff and puff process. In this Bakken model, because of the large unstimulated area, a substantial amount of CO_2 did not arrive at the producer so that most of CO_2 was stored in the reservoir. In the case of CO_2 huff and puff, however, CO_2 could not spread to the unstimulated zone owing to the injection, soaking, and production cycles were performed in one short step. Therefore, the CO_2 flooding method is more effective than the CO_2 huff and puff process, both in oil production and CO_2 storage, in the Bakken oil reservoir.

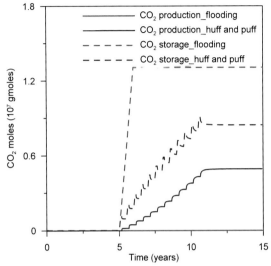

FIG. 5.24 CO_2 production and storage of the CO_2 flooding and huff and puff cases in the Bakken models.

To consider diffusive transport between hydrocarbon components and CO_2 in shale oil reservoirs, molecular diffusion was considered in the model. The Wilke-Chang correlation was used to simulate the molecular diffusion mechanism in the Bakken oil model. Fig. 5.25 shows distributions of CO_2 mole fractions for the CO_2 flooding, depending on the consideration of molecular diffusion in the fracture and matrix grids after 2 years of CO_2 injection. Figs. 5.25A and B show fracture and matrix grids of models without molecular diffusion, and Fig. 5.25C shows a model with molecular diffusion. As shown in Fig. 5.25A, CO_2 flows excessively through the fracture grids without consideration of molecular diffusion. In the matrix, there is only a slight flow near the injector, and CO_2 cannot flow to the producer (Fig. 5.25B). When Wilke-Chang molecular diffusion is considered, CO_2 is smoothly transported as shown in Fig. 5.25C. In the model with molecular diffusion, the distribution of CO_2 is almost same in both the fracture and matrix grids. The model without molecular diffusion (Fig. 5.25A) shows more rapid CO_2 flow than the model with molecular diffusion (Fig. 5.25C) in the fracture grids. In the model without Wilke-Chang correlation, diffusion through the matrix is neglected and, as a result, oil recovery remains low. The results indicate that the efficiency of CO_2 flooding increases by molecular diffusion in the low

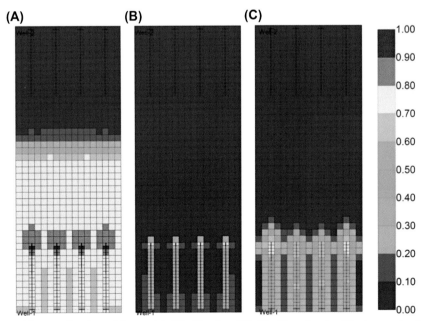

FIG. 5.25 CO_2 mole fractions of the models **(A)** without and **(B, C)** with molecular diffusion mechanism in the **(A, B)** fracture and **(A, C)** matrix grids after 2 years of CO_2 flooding.

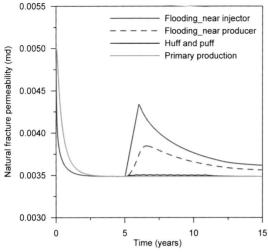

FIG. 5.26 Natural fracture permeability variation of the CO_2 flooding, huff and puff, and no injection cases in the Bakken models.

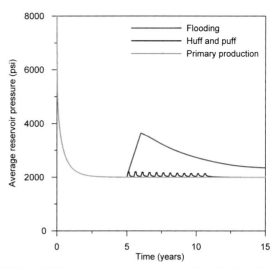

FIG. 5.27 Average reservoir pressure of the CO_2 flooding, huff and puff, and no injection cases in the Bakken models.

permeability matrix. Because of the low permeability of shale oil reservoir, the effects of diffusion are more significant than those in conventional reservoirs. When simulating CO_2 injection into shale oil reservoirs, molecular diffusion mechanisms should be considered.

The rock and fracture properties of shale reservoirs are sensitive to variations of stress and strain during production and injection. To consider the variation of these properties, stress-dependent deformation should be simulated in the shale reservoir model (Cho et al., 2013). Exponential correlations were used to calculate the stress-dependent deformation in Bakken model. Fig. 5.26 shows the change of the natural fracture permeability for CO_2 flooding, huff and puff, and no injection cases. In the primary production period, natural fracture permeability decreases rapidly by about 30% due to the pressure decrease. After CO_2 injection starts, natural fracture permeability increases during the injection period, and then decreases again. In the case of CO_2 flooding, the average pressure increases (Fig. 5.27) so that the natural fracture permeability increases (Fig. 5.26) due to the pressurizing effect. This effect is larger in the region near the injection well than the region near the production well. The increase of natural fracture permeability of the CO_2 huff and puff case is lower than that of the CO_2 flooding case because most of injected CO_2 is reproduced just after injection. The increase of natural fracture permeability caused by stress-dependent deformation has a positive effect on oil production. Especially, these results show that the

geomechanic effects are remarkable with the CO_2 flooding process in the Bakken reservoir.

To analyze the effects of a horizontal well and hydraulic fractures in a real field situation, the numerical model coupled with fracking simulator was constructed. Based on logging and geologic data of real field, hydraulic fractures are simulated by GOHFER (Barree & Associates, 2015). Five horizontal wells with six fracturing stages are drilled in this model. Four perforations are allowed per stage, and perforation spacing is 150 ft. As shown in Fig. 5.28, a shale oil reservoir is simulated by setting up a heterogeneous reservoir model with dimensions of 3550 ft×5150 × 490 ft. Initial reservoir pressure and temperature are 4050 psi and 250°F. Ranges of permeability and porosity of reservoir are 0.001−0.028 md, 0.047−0.112, respectively. Generated hydraulic fractures show 160−320 ft of half-length.

In this shale oil reservoir, the effects of CO_2 flooding were investigated. Owing to the extremely high computational cost of the model, segments of this shale oil reservoir was considered with three horizontal wells with hydraulic fractures. To investigate the effects of different horizontal wells, hydraulic fractures, and reservoir properties, six models were simulated. Models 1 and 2 include well 1, 2, and 3. Models 3 and 4 include well 2, 3, and 4. Models 5 and 6 include well 3, 4, and 5. Models 1, 3, and 5 are the upper part of the reservoir and Models 2, 4, and 6 are the lower part of the reservoir. For example, Models 1 and 2 are marked with red and yellow lines, respectively, in Fig. 5.28. During

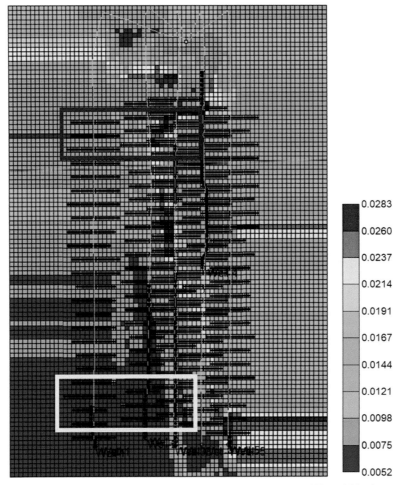

FIG. 5.28 Reservoir model of shale oil reservoir with permeability grid (md).

the primary production period in the first year, every horizontal well was produced. After the primary production period, CO_2 is injected into the middle horizontal well in each model while side horizontal wells produce continuously for a year.

Figs. 5.29–5.31 present oil recovery of the model with and without CO_2 injection cases in the Models 1 through 6. As shown in these results, oil recoveries in Models 1 through 6 are improved by 21%, 6%, 29%, 18%, 9%, and 28%, respectively. Owing to the connectivity of hydraulic fractures in these models, oil improvement is observed immediately after CO_2 injection in most models. Among these models, however, oil improvement of Model 2 is observed slowly, and it shows the lowest improvement. In Model 2, the spacing between wells 1 and 2 is 1100 ft, which is the longest in this model. Hydraulic fracture half-length of well 1 in

Model 2 is short compared with another part so that Model 2 shows low connectivity between hydraulic fractures. In addition, reservoir permeability is also small in this model as shown in Fig. 5.28. As shown in Fig. 5.32, the effect of CO_2 injection in well 1 is insignificant compared with well 3. As shown in Fig. 5.31A, Model 5 also shows low improvement of oil recovery. Model 5 indicates the upper part including wells 3, 4, and 5. In contrast with Model 2, Model 5 shows high connectivity of hydraulic fractures and high matrix permeability. The reason for low oil recovery in this model is low hydraulic fracture permeability. Hydraulic fracturing process was not well performed because of the geomechanic problem in this region so that permeability of hydraulic fractures is lower than other fractures. Therefore, for field application of CO_2 injection in shale reservoirs, accurate understanding of horizontal

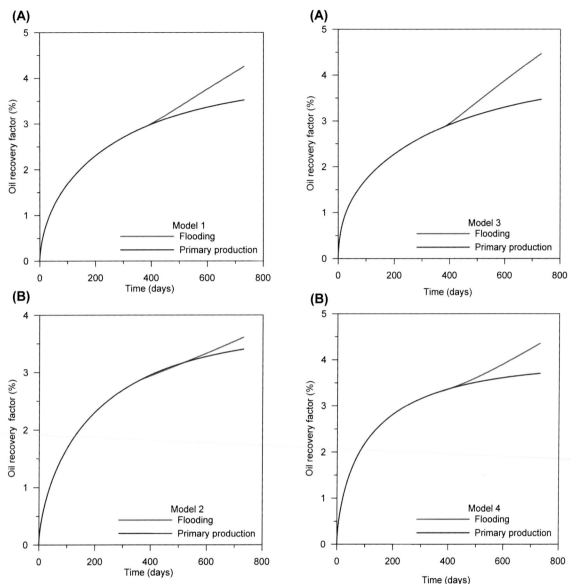

FIG. 5.29 Oil recovery of the model with and without CO_2 injection in the Models **(A)** 1 and **(B)** 2.

FIG. 5.30 Oil recovery of the model with and without CO_2 injection in the Models **(A)** 3 and **(B)** 4.

well and hydraulic fractures as well as geological properties and various specific mechanisms are significantly important.

CONSIDERATION OF ORGANIC MATERIAL IN SHALE RESERVOIR SIMULATION

In the petroleum industry, various simulations on shale oil and gas reservoirs have been performed. The main concerns of these models include low permeability

and low porosity of the matrix, permeability and spacing of natural fracture systems, and height, width, spacing, half-length, and permeability of hydraulic fractures. Specific mechanisms such as adsorption, non-Darcy flow, and stress-dependent deformation are also considered these days. It is commonly known that accurate characterization of hydraulic fracture is the most important in the shale reservoirs so that many oil corporations have focused mainly on the analysis of

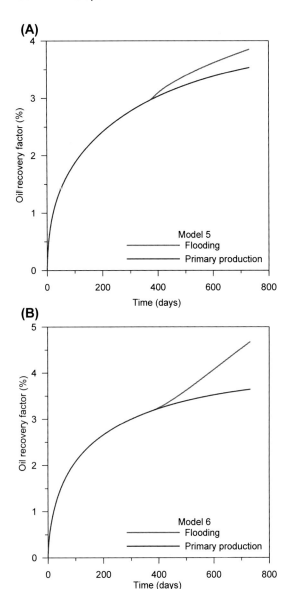

(A)

(B)

FIG. 5.31 Oil recovery of the model with and without CO₂ injection in the Models **(A)** 5 and **(B)** 6.

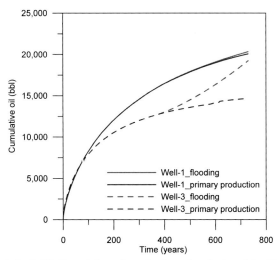

FIG. 5.32 Cumulative oil production of wells 1 and 3 with and without CO_2 injection in Model 2.

hydraulic fractures. Numerical models including these properties and mechanisms have presented proper results up to a certain point. However, as suggested in Chapters 2 and 3, unconventional shale reservoirs are more complex than these industry reservoir models. Flow mechanisms from nano-to macro-scale system of shale reservoirs should be fully considered as shown in Fig. 5.33 (Zhang et al., 2017). Especially, effects of organic matters in shale reservoirs attract significant interest in recent years.

Several researchers presented that the pore structure of the shale matrix is significantly involved, and it is comprised of organic pores such as kerogen and bitumen and inorganic pores (Ko, Loucks, Ruppel, Zhang, & Peng, 2017, 2016; Loucks, Reed, Ruppel, & Hammes, 2012; Pommer & Milliken, 2015; Zhang et al., 2017). Generally, the pores in organic matters are in nanoscale, and the inorganic pores range from nanoscale to microscale. Thus, dominant transport mechanisms in organic pores and inorganic pores can be different. Depending on a recent study (Hao, Adwait, Hussein, Xundan, & Lin, 2015), the adsorption and desorption principally occur not in the inorganic matters, but in the organic materials. According to experiments for organic-rich shale cores performed by Akkutlu and Fathi (2012), desorption, diffusion, and dissolution in organic materials dominate the process of gas flux in shale cores. Nanopores in organic matters provide abundant space for storage and transport of hydrocarbons. In addition, porosity and permeability of organic pores are different from inorganic pores. Geomechanic properties of organic materials are also different from those of inorganic materials. TOC in shale reservoirs are at least 2%, and it can exceed 10%–12%. Therefore, effects of organic matters in shale reservoirs cannot be ignored, and it would be more reasonable to distinguish the organic matters from inorganic materials for accurate modeling of shale reservoirs. Because most conventional simulators use the simple matrix models not considering organic material these days, they are inadequate to explain the fluid flow

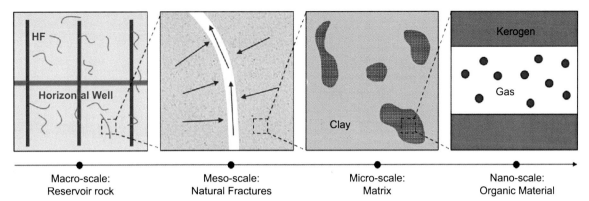

FIG. 5.33 Fluid flow from nano-to macro-scale system of shale reservoirs. (Credit: Zhang, W., J. Xu, R. Jiang, Y. Cui, J. Qiao, C. Kang, et al. (2017). Employing a quad-porosity numerical model to analyze the productivity of shale gas reservoir. *Journal of Petroleum Science and Engineering, 157*, 1046–1055. doi:10.1016/j.petrol. 2017.07.031.)

in the shale matrix (Li, Xu, Wang, & Lu, 2014). In the following section, effects of organic materials in shale reservoirs are presented.

Pore system of organic matters and inorganic materials are significantly different. There are numerous studies to evaluate and distinguish the organic and inorganic pore properties (Loucks et al., 2012; Bernard et al., 2012a, 2012b; Milliken, Rudnicki, Awwiller, & Zhang, 2013; Löhr, Baruch, Hall, & Kennedy, 2015). Because of diversity and uncertainty of pore system of shale reservoirs, there is still no consensus about it. Generally, porosity in the organic and inorganic material is highly dependent on the maturity of reservoirs. As thermal maturity increases, inorganic, mineral-associated porosity decreases due to diagenesis, such as compaction, cementation, and bitumen infill (Hu et al., 2017). The size of organic matter pores and porosity first increase with thermal maturity because hydrocarbons are generated and expelled from the organic materials, and secondary cracking occurred (Modica & Lapierre, 2012; Milliken et al., 2013; Curtis et al., 2012; Pommer & Milliken, 2015; Hu et al., 2017). In the late postmature stage, organic matter porosity decreases with increasing of maturity due to consequent pore collapse by overmatured condition (Milliken et al., 2013; Hu et al., 2017). Fig. 5.34 shows mineral-associated and organic matter porosities depending on vitrinite reflectance in Marcellus Shale, Eagle Ford Shale, and Wufeng-Longmaxi Shale (Hu et al., 2017). Mastalerz, Schimmelmann, Drobniak, and Chen (2013) also presented that relative properties of micropores, mesopores, and macropores within the New Albany Shales changed with increasing thermal maturity levels owing to hydrocarbon generation and migration. Hu

et al. (2015) showed that mesopores, especially pores of 2–6 nm in diameter, increased with thermal maturity levels in artificial maturation system. In the organic matters, pore-dominant reservoirs such as Barnett Shale, Marcellus Shale, Woodford Shale, and Wufeng-Longmaxi Shale (Hu et al., 2017; Löhr et al., 2015; Loucks, ReedRuppel, & Jarvie, 2009; Milliken et al., 2013), organic matter hosted porosity and pore size increased and then decreased with the increase of TOC.

Depending on Ross and Bustin (2009), adsorption capacity increases as TOC increases and moisture content decreases. Water molecules are adsorbed on specific hydrophilic sites, and methane molecules are adsorbed on other available sorption sites. Generally, clay minerals, which present hydrophilic nature, reduce hydrocarbon adsorption capacity, and organic matters, which have hydrophobic nature, provide sites for hydrocarbon adsorption. Several previous studies observed that methane adsorption is positively correlated with TOC in shale reservoir (Lu, Li, & Watson, 1995; Ross & Bustin, 2007; Cui, Bustin, & Bustin, 2009). Strapoc, Mastalerz, Schimmelmann, Drobniak, and Hasenmueller (2010) presented that a positive correlation between total gas content from desorption of shale cores and organic matter content is mainly responsible for total gas-in-place in New Albany Shale. Ross and Bustin (2009) observed that methane adsorption increases as TOC and micropore volume, which is less than 2 nm, increase. The results indicated that microporosity, which is dominant in organic matters, is the main factor for methane adsorption. Zhang, Ellis, Ruppel, Milliken, and Yang (2012) presented methane adsorption characteristics of bulk shale core and kerogen isolated from bulk shale core. Numerous

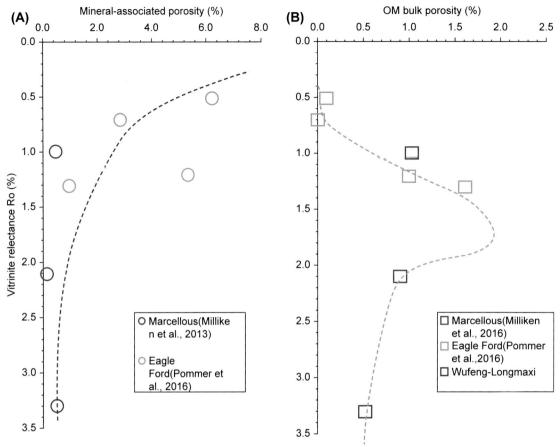

FIG. 5.34 Mineral-associated and organic matter porosities depending on vitrinite reflectance in Marcellus, Eagle Ford, and Wufeng-Longmaxi Shale. (Credit: Hu, H. Y., Hao, F., Lin, J. F., Lu, Y. C., Ma, Y. Q., & Li, Q. (2017). Organic matter-hosted pore system in the Wufeng-Longmaxi (O(3)w-S(1)1) shale, Jiaoshiba area, Eastern Sichuan Basin, China. *International Journal of Coal Geology, 173*, 40–50. doi:10.1016/j.coal.2017.02. 004.)

methane adsorption experiments on the two samples were performed at different temperatures. Fig. 5.35 shows that comparison between organic-rich shale core and its isolated kerogen (Zhang et al., 2012). Amount of methane adsorption in isolated kerogen is larger than the bulk core sample. These results indicate that methane adsorption is performed only on organic matters. In both samples of bulk and isolated kerogen, the amount of methane adsorption is proportional to TOC. The results present that adsorption and desorption on organic materials play an important role in the storage of shale reservoirs. In other words, the capacity of gas adsorption increases as TOC content increases. In addition, kerogen types affect Langmuir pressure, which is the pressure corresponding to one-half Langmuir volume. Thermal maturity influences the capacity

of gas adsorption of organic-rich shale at the low-pressure condition. Therefore, adsorption/desorption mechanisms should be considered in organic matters separated from inorganic materials.

Geomechanic properties of organic matters are also different from inorganic materials. Owing to the abundant presence, organic matters could considerably affect the mechanical behavior of shale rocks. However, studies on the mechanical properties of organic matters still lack these days. Aoudia, Miskimins, Harris, and Mnich (2010) and Kumar, Sondergeld, and Rai (2012) provided total organic carbon is inversely proportional to Young's modulus of shale rock. Depending on Han, Al-Muntasheri, Katherine, and Abousleiman (2016), the tensile strength of kerogen is an order of magnitude higher than shales. Its tensile behavior

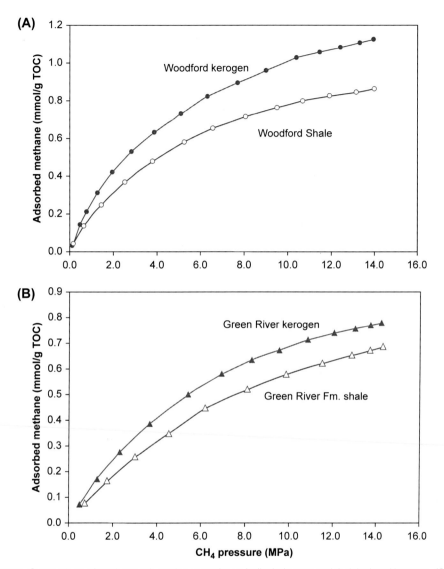

FIG. 5.35 Comparison of methane adsorption capacity on bulk shale core and their isolated kerogen. (Credit: Zhang, T., Ellis, G.S., Ruppel, S.C., Milliken, K., & Yang, R. (2012). Effect of organic-matter type and thermal maturity on methane adsorption in shale-gas systems. *Organic Geochemistry, 47*, 120–131. doi:10.1016/j. orggeochem.2012.03.012.)

demonstrates strain softening characteristics (Fig. 5.36), which have serious implications for hydraulic fracturing in kerogen-rich shales. Khatibi et al. (2018) presented a method to measure the mechanical properties of organic matters using Raman spectroscopy, which is a function of molecular structure and chemical compounds. The results presented that organic material is the least stiff constitute in shale rock. Organic matters in shale show strengthening effect so that higher injection pressure is needed or hydraulic fracturing in organic-rich shale reservoirs. The strain-softening behavior of organic matters under loading condition implies that the tensile strength along the fracture of the organic-rich shale does not decrease as the inorganic

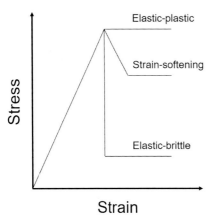

FIG. 5.36 Possible failure mechanisms of a rock mass.

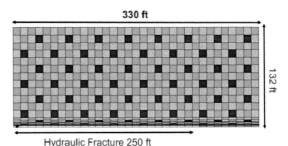

FIG. 5.37 Schematic views of shale reservoir model considering different porosity systems. (Credit: Khoshghadam, M., Khanal, A., Rabinejadganji, N., & John Lee, W. (2016). How to model and improve our understanding of liquid-rich shale reservoirs with complex organic/inorganic pore network. *Paper presented at the SPE/AAPG/SEG unconventional resources technology conference, San Antonio, Texas, USA, 2016/2018/1.* doi:10.15530/URTEC-2016-2459272.)

shale. For accurate modeling of shale reservoirs, mechanical properties of organic matters also should be considered.

Although several imaging measurements have revealed that shale pore networks are comprised of organic matter, inorganic materials, and natural fractures; research for flow mechanisms considering these complex geometry is still in an early stage. Simulation studies examining the effects of complex geometry including organic matters in shale reservoirs have been introduced recently. Khoshghadam, Khanal, Rabinejadganji, and John Lee (2016) conducted comprehensive simulation studies considering four different porosity systems indicating distinctive characteristics such as organic and inorganic materials in the shale matrix and natural and hydraulic fractures in liquid-rich shale reservoirs. They insisted that the connectivity of four pore systems and relative permeability are significant and uncertainty of them is critical for the simulation of fluid flow in liquid-rich shale reservoirs.

As shown in Fig. 5.37, Khoshghadam et al. (2016) divided the shale matrix into organic matter pores and inorganic materials. Green and red blocks indicate organic matter pores, and gray blocks indicate inorganic pores. They further divided organic matters into pores smaller than 10 nm (green blocks) and larger than 10 nm (red blocks). In organic and inorganic materials, properties such as porosity, permeability, initial water saturation, and compressibility are different. Khoshghadam et al. (2016) assumed that organic matters have higher porosity and lower permeability compared with inorganic materials based on the basic nature of pores (Honarpour, Nagarajan, Orangi, Arasteh, & Yao, 2012). Relationship of permeability-porosity-pore throat size shows that permeability is

proportional to the porosity multiplied by the square of pore throat size (Nelson & Batzle, 2006). Based on the Carman-Kozeny equation (Kozeny, 1927; Carman, 1937, 1956), permeability increases as pore size increases. Various studies presented that every shale reservoir has specific properties so that an in-depth study for geological properties of shale reservoir is required for the accurate model.

Khoshghadam et al. (2016) also considered the enhancement of critical gas saturation. Liberation of a gas molecule from oil is composed of some processes such as nucleation, bubble growth, coalescence, and bulk gas phase formation (Honarpour et al., 2012). To create a stable path through a liquid phase, critical gas saturation should be reached. Critical gas saturation is small in high-permeability formations because pores size is large, and the fraction of pores needed to be filled with gas molecules is smaller (Chu, Ye, Harmawan, Du, & Shepard, 2012). Permeability and pore size are small in nanopores. Therefore, gas bubbles have large curvature due to small pore size so that more substantial pressure difference is required between liquid and gas phases. The gas phase should reach a higher saturation to sustain stable and movable gas bubbles. Generally, organic matter pores are mostly oil-wet with low initial water saturation, whereas inorganic pores are water-wet or mixed-wet pores with high initial water saturation. Even if inorganic pores often have heterogeneously mixed wettability, Khoshghadam et al. (2016) assumed that inorganic pores are water-wet for simplicity. Because of the small pore size of organic matter pores, high values of critical gas saturation, residual oil saturation, and Corey exponent of organic pore grids are used compared with those of inorganic pore grids Khoshghadam et al. (2016) presented that relative permeability functions and critical gas saturation are essential parameters in multiphase liquid-rich shale oil reservoirs. Although critical gas saturation and relative permeability may not be important compared with other factors in the early stages of production, they will play an essential role later. It is because, in an early stage, the reservoirs pressure is higher than bubble point and reservoirs behave as single-phase flow and, subsequently, the reservoir pressure decreases below saturation pressure and reservoirs become two phases.

In addition, Khoshghadam et al. (2016) considered confinement effects on phase behavior in organic nanopores, stimulated reservoir volume with changing natural fracture permeability depending on the distance from the hydraulic fracture, and rock compaction. Simulation results revealed that hydrocarbon production from liquid-rich shale reservoirs exhibits complex dynamics relying on the complex pore network, thermal maturity level, volatility of the reservoir fluid, and hydraulic fracturing. However, in their study, the results presented that organic matter pores have little effect on the production of liquid-rich shale reservoirs. For accurate modeling of organic matters in shale reservoirs, more detailed information of organic materials is required. As mentioned earlier, adsorption capacity and geomechanic properties, as well as hydraulic properties of organic matters, should be explicitly investigated and applied.

As mentioned in Chapter 3, gas transport in organic materials deviates from Darcy flow. In micro- and nanoscale pore matrix of shale reservoirs, Darcy's law cannot adequately replicate the gas flow. With small pore size comparable with the mean free path, Knudsen diffusion becomes dominant. On the other hand, at large pores such as inorganic pore and fractures, Klinkenberg slip flow is governing flow mechanism. An apparent permeability considering gas transport mechanisms such as Knudsen flow, flip flow, and gas adsorption in nanopore matrix were presented by several authors (Azom & Javadpour, 2012; Darabi, Ettehad, Javadpour, & Sepehrnoori, 2012; Javadpour, Fisher, & Unsworth, 2007; Javadpour, 2009; Singh, Javadpour, Ettehadtavakkol, & Darabi, 2014; Singh & Javadpour, 2016). Yuan et al. (2014) investigated the effects of methane storage and diffusion in shale rock experiments. They presented that main transport mechanisms in macropores and micropores are Fickian diffusion and Knudsen diffusion, respectively. Sheng et al. (2015) and Pang, Soliman, Deng, and Xie (2017) also presented that surface diffusion on slippage is primarily dominated by the diffusion coefficient and pore size.

Various studies were performed to develop an analytical model of fluid flow in shale reservoirs. Table 5.3 presents comparisons of previous analytical flow models in shale reservoirs (Fan & Ettehadtavakkol, 2017). In Table 5.3, PSS and TR indicate psedosteady state and transient state, respectively. Fan and Ettehadtavakkol (2017) developed the analytic model with comprehensive transport mechanisms in complex pore systems including organic matters, inorganic

---

TABLE 5.3
Comparisons of Analytical Flow Models in Shale Reservoirs

Author	Diffusion in SRV	Diffusion in Organic Matters	Desorption	Geomechanics
Stalgorova and Mattar (2013)	Normal	—	—	—
Wang, Shahvali, and Su (2015)	Anomalous	—	—	—
Albinali and Ozkan (2016)	Anomalous	—	—	—
Tabatabaie, Pooladi-Darvish, Mattar, and Tavallali (2017)	Normal	—	—	Exponential
Chen, Liao, Zhao, Dou, and Zhu (2016)	Normal	PSS	Langmuir	Exponential
Fan and Ettehadtavakkol (2017)	Anomalous	TR	Langmuir	Exponential

Reproduced from Fan, D., & Ettehadtavakkol, A. (2017). Analytical model of gas transport in heterogeneous hydraulically-fractured organic-rich shale media, *Fuel, 207,* 625–640. doi:10.1016/j.fuel.2017.06.105.

clay, induced fracture, and hydraulic fracture. They considered gas diffusion and desorption in organic matters, slip flow in inorganic materials and fractures, transient Darcy flow from the matrix to induced fractures, Darcy flow in both induced and hydraulic fractures, tortuosity in induced fractures, and pressure-dependent permeability. In addition, they applied their analytic model to perform history matching and prediction processes at the field scale.

Fig. 5.38 presents schematic views of flow in analytic shale reservoir model (Fan & Ettehadtavakkol, 2017). In the overall reservoir, the flow region is divided into the hydraulic fracture (HF), stimulated reservoir volume (SRV), and unstimulated reservoir volume (USRV) depending on flow conductivity (Fig. 5.38A). In SRV, the bulk matrix is divided by the organic matrix, the organic pores, the inorganic matrix, and the inorganic

pores (Fig. 5.38C). In the organic matrix and pores, diffusion and desorption mechanisms are dominant. Governing diffusivity equations in organic matrix and organic pores are given as:

$$\frac{\partial^2 C_{kD}}{\partial z_D'^2} = \lambda_k \frac{\partial C_{kD}}{\partial t_D}, \tag{5.1}$$

$$\frac{\partial^2 C_{pD}}{\partial y_D'^2} + \frac{1}{h_{kD}^2} \frac{\partial^2 C_{pD}}{\partial z_D'^2} = \lambda_p \frac{\partial C_{pD}}{\partial t_D}, \tag{5.2}$$

where C is the concentration, y' and z' are the coordinate in organic matter, λ is the interporosity flow coefficient, and h is the thickness. Subscripts of p, k, and D indicate organic pores, organic matrix (kerogen), and dimensionless, respectively. Depending on the gas concentration gradient, dissolved gas in the organic matrix diffuses to the subsurface of organic matter pores. Fick's

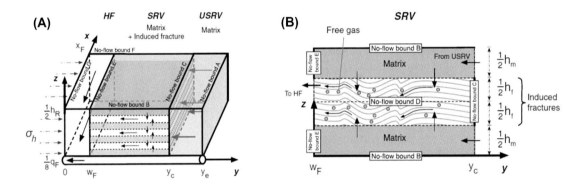

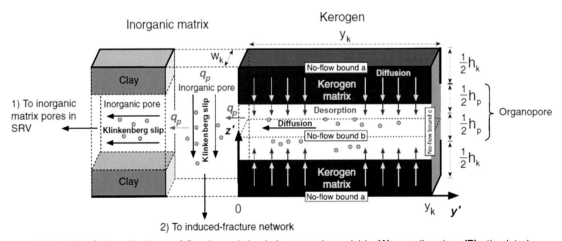

FIG. 5.38 Schematic views of flow in analytic shale reservoir model in **(A)** overall region, **(B)** stimulated reservoir volume (SRV), and **(C)** organic-rich matrix. (Credit: Fan, D., & Ettehadtavakkol, A. (2017). Analytical model of gas transport in heterogeneous hydraulically-fractured organic-rich shale media. *Fuel*, *207*, 625–640. doi:10.1016/j.fuel.2017.06.105.)

second law is used to compute transient diffusion in organic matrix and pores. If pore pressure decreases during production, adsorbed gas on the organic matrix start to desorb. Governing diffusivity equations in USRV, matrix in SRV, induced fracture network in SRV, and HF were presented as follows:

$$\frac{\partial^2 \psi_{UD}}{\partial \gamma_D^2} - \gamma_D \left(\frac{\partial \psi_{UD}}{\partial \gamma_D} \right)^2 = \omega_f e^{\gamma_D \psi_{UD}}$$
$$\left(\frac{1}{\eta_{mD1}} \frac{\partial \psi_{UD}}{\partial t_D} - \frac{1}{\eta_{mD2}} \frac{1-\omega_m}{\lambda_p \omega_m} \frac{\partial C_{pD}}{\partial \gamma'_D} \bigg|_{\gamma'_D = 0} \right), \tag{5.3}$$

$$\frac{\partial^2 \psi_{mD}}{\partial z_D^2} + h_{mD}^2 \frac{\partial^2 \psi_{mD}}{\partial \gamma_D^2} - \gamma_D \left[\left(\frac{\partial \psi_{mD}}{\partial z_D} \right)^2 + h_{mD}^2 \left(\frac{\partial \psi_{mD}}{\partial \gamma_D} \right)^2 \right]$$
$$= \omega_f h_{mD}^2 e^{\gamma_D \psi_{mD}} \left(\frac{1}{\eta_{mD1}} \frac{\partial \psi_{mD}}{\partial t_D} - \frac{1}{\eta_{mD2}} \frac{1-\omega_m}{\lambda_p \omega_m} \frac{\partial C_{pD}}{\partial \gamma'_D} \bigg|_{\gamma'_D = 0} \right), \tag{5.4}$$

$$\frac{\partial^2 \psi_{fD}}{\partial \gamma_D^2} + \frac{1}{h_{mD}^2} \frac{\partial^2 \psi_{fD}}{\partial z_D^2} - \frac{\theta}{\gamma_D} \frac{\partial \psi_{fD}}{\partial \gamma_D} - \gamma_D \left[\left(\frac{\partial \psi_{fD}}{\partial \gamma_D} \right)^2 + \frac{1}{h_{mD}^2} \left(\frac{\partial \psi_{fD}}{\partial z_D} \right)^2 \right]$$
$$= w_D^{-\theta} \omega_f e^{\gamma_D \psi_{fD}} \gamma_D^\theta \frac{\partial \psi_{fD}}{\partial t_D}, \tag{5.5}$$

$$\frac{\partial^2 \psi_{FD}}{\partial x_D^2} + \frac{\partial^2 \psi_{FD}}{\partial \gamma_D^2} - \gamma_D \left[\left(\frac{\partial \psi_{FD}}{\partial x_D} \right)^2 + \left(\frac{\partial \psi_{FD}}{\partial \gamma_D} \right)^2 \right]$$
$$= \frac{\omega_f}{\eta_{FD}} e^{\gamma_D \psi_{FD}} \frac{\partial \psi_{FD}}{\partial t_D}, \tag{5.6}$$

where ψ is the pseudopressure, γ the permeability modulus, ω the storativity ratio, η the diffusivity, θ the tortuosity index of the induced-fracture network, w the weight fraction of clay, and x_D, γ_D, and z_D are the coordinates in the reservoir model. Subscripts U, f, m, and F indicate USRV matrix, induced fracture, inorganic matrix, and hydraulic fracture, respectively. In USRV, gas first diffuses and desorbs in organic matters, and then slips along inorganic matrix pores. In the matrix of SRV, the gas flow is composed of flux from the organic matrix in matrix of SRV and total flow from USRV. In induced fracture of SRV, diffusivity is a function of pressure-dependent permeability and tortuosity index. Details of these gas flow mechanisms and calculation of equations are presented by Fan and Ettehadtavakkol (2017).

REFERENCES

Akkutlu, I. Y., & Fathi, E. (2012). Multiscale gas transport in shales with local Kerogen heterogeneities. *SPE Journal*, 17(04), 1002−1011. https://doi.org/10.2118/146422-PA.

Albinali, A., & Ozkan, E. (2016). Anomalous diffusion approach and field application for fractured nano-porous reservoirs. In *Paper presented at the SPE annual technical conference and exhibition, Dubai, UAE, 2016/9/26*. https://doi.org/10.2118/181255-MS.

Alharthy, N., Teklu, T., Kazemi, H., Graves, R., Hawthorne, S., Braunberger, J., et al. (2015). Enhanced oil recovery in liquid-rich shale reservoirs: Laboratory to field. In *Paper presented at the SPE annual technical conference and exhibition, Houston, Texas, USA, 2015/9/28*. https://doi.org/10.2118/175034-MS.

Aoudia, K., Miskimins, J. L., Harris, N. B., & Mnich, C. A. (2010). Statistical analysis of the effects of mineralogy on rock mechanical properties of the Woodford Shale and the associated impacts for hydraulic fracture treatment design. In *Paper presented at the 44th U.S. Rock Mechanics symposium and 5th U.S.-Canada rock Mechanics symposium, Salt Lake city, Utah, 2010/1/1*.

Azom, P. N., & Javadpour, F. (2012). Dual-continuum modeling of shale and tight gas reservoirs. In *Paper presented at the SPE annual technical conference and exhibition, San antonio, Texas, USA, 2012/1/1*. https://doi.org/10.2118/159584-MS.

Barree & Associates. (2015). *Gohfer user manual*. Barree & Associates LLC.

Bernard, S., Horsfield, B., Schulz, H.-M., Wirth, R., Schreiber, A., & Sherwood, N. (2012a). Geochemical evolution of organic-rich shales with increasing maturity: A STXM and TEM study of the posidonia shale (lower toarcian, Northern Germany). *Marine and Petroleum Geology*, 31(1), 70−89. https://doi.org/10.1016/j.marpetgeo.2011.05.010.

Bernard, S., Wirth, R., Schreiber, A., Schulz, H.-M., & Horsfield, B. (2012b). Formation of nanoporous pyrobitumen residues during maturation of the Barnett shale (Fort Worth Basin). *International Journal of Coal Geology*, 103, 3−11. https://doi.org/10.1016/j.coal.2012.04.010.

Busch, A., Alles, S., Gensterblum, Y., Prinz, D., Dewhurst, D. N., Raven, M. D., et al. (2008). Carbon dioxide storage potential of shales. *International Journal of Greenhouse Gas Control*, 2(3), 297−308. https://doi.org/10.1016/j.ijggc.2008.03.003.

Carman, P. C. (1937). Fluid flow through granular beds. *Transactions of the Institution of Chemical Engineers*, 15, 150−166.

Carman, P. C. (1956). *Flow of gases through porous media*. Academic Press.

Chen, C., Balhoff, M. T., & Mohanty, K. K. (2014). Effect of reservoir heterogeneity on primary recovery and CO₂ huff 'n' puff recovery in shale-oil reservoirs. *SPE Reservoir Evaluation & Engineering*, 17(03), 404−413. https://doi.org/10.2118/164553-PA.

Chen, Z., Liao, X., Zhao, X., Dou, X., & Zhu, L. (2016). Development of a trilinear-flow model for carbon sequestration in depleted shale. *SPE Journal*, 21(04), 1386−1399. https://doi.org/10.2118/176153-PA.

Cho, Y., Ozkan, E., & Apaydin, O. G. (2013). Pressure-dependent natural-fracture permeability in shale and its effect on shale-gas well production. *SPE Reservoir Evaluation & Engineering*, 16(02), 216−228. https://doi.org/10.2118/159801-PA.

Chu, L., Ye, P., Harmawan, I. S., Du, L., & Shepard, L. R. (2012). Characterizing and simulating the nonstationariness and nonlinearity in unconventional oil reservoirs: Bakken application. In *Paper presented at the SPE Canadian unconventional resources conference, Calgary, Alberta, Canada, 2012/1/1.* https://doi.org/10.2118/161137-MS.

Cui, X., Bustin, A. M. M., & Bustin, R. M. (2009). Measurements of gas permeability and diffusivity of tight reservoir rocks: Different approaches and their applications. *Geofluids, 9*(3), 208–223. https://doi.org/10.1111/j.1468-8123.2009.00244.x.

Curtis, M. E., Cardott, B. J., Sondergeld, C. H., & Rai, C. S. (2012). Development of organic porosity in the Woodford Shale with increasing thermal maturity. *International Journal of Coal Geology, 103,* 26 31. https://doi.org/10.1016/j.coal.2012.08.004.

Darabi, H., Ettehad, A., Javadpour, F., & Sepehrnoori, K. (2012). Gas flow in ultra-tight shale strata. *Journal of Fluid Mechanics, 710,* 641–658. https://doi.org/10.1017/jfm.2012.424.

Eshkalak, M. O., Al-shalabi, E. W., Sanaei, A., Aybar, U., & Sepehrnoori, K. (2014). Enhanced gas recovery by CO_2 sequestration versus Re-fracturing treatment in unconventional shale gas reservoirs. In *Paper presented at the Abu Dhabi international petroleum exhibition and conference, Abu Dhabi, UAE, 2014/11/10.* https://doi.org/10.2118/172083-MS.

Fan, D., & Ettehadtavakkol, A. (2017). Analytical model of gas transport in heterogeneous hydraulically-fractured organic-rich shale media. *Fuel, 207,* 625–640. https://doi.org/10.1016/j.fuel.2017.06.105.

Fathi, E., & Akkutlu, I. Y. (2014). Multi-component gas transport and adsorption effects during CO_2 injection and enhanced shale gas recovery. *International Journal of Coal Geology, 123,* 52–61. https://doi.org/10.1016/j.coal.2013.07.021.

Gamadi, T. D., Sheng, J. J., & Soliman, M. Y. (2013). An experimental study of cyclic gas injection to improve shale oil recovery. In *Paper presented at the SPE annual technical conference and exhibition, New Orleans, Louisiana, USA, 2013/2019/30.* https://doi.org/10.2118/166334-MS.

Gamadi, T. D., Sheng, J. J., Soliman, M. Y., Menouar, H., Watson, M. C., & Emadibaladehi, H. (2014). An experimental study of cyclic CO_2 injection to improve shale oil recovery. In *Paper presented at the SPE improved oil recovery symposium, Tulsa, Oklahoma, USA, 2014/4/12.* https://doi.org/10.2118/169142-MS.

Ghorbae, S. Z., & Alkhansa, Z. (2012). Evaluation of the effects of molecular diffusion in recovery from fractured reservoirs during gas injection. In *Paper presented at the SPE Kuwait international petroleum conference and exhibition, Kuwait City, Kuwait, 2012/1/1.* https://doi.org/10.2118/163273-MS.

Godec, M., Koperna, G., Petrusak, R., & Oudinot, A. (2013). Potential for enhanced gas recovery and CO_2 storage in the Marcellus shale in the Eastern United States. *International Journal of Coal Geology, 118,* 95–104. https://doi.org/10.1016/j.coal.2013.05.007.

Godec, M., Koperna, G., Petrusak, R., & Oudinot, A. (2014). Enhanced gas recovery and CO_2 storage in gas shales: A summary review of its status and potential. *Energy Procedia, 63,* 5849–5857. https://doi.org/10.1016/j.egypro.2014.11.618.

Han, Y., Al-Muntasheri, G., Katherine, L. H., & Abousleiman, Y. N. (2016). Tensile mechanical behavior of kerogen and its potential implication to fracture opening in Kerogen-Rich Shales (KRS). In *Paper presented at the 50th U.S. Rock Mechanics/geomechanics symposium, Houston, Texas, 2016/6/26.*

Hao, S., Adwait, C., Hussein, H., Xundan, S., & Lin, L. (2015). Understanding shale gas flow behavior using numerical simulation. *SPE Journal, 20*(01), 142–154. https://doi.org/10.2118/167753-PA.

Hawthorne, S. B., Gorecki, C. D., Sorensen, J. A., Steadman, E. N., Harju, J. A., & Melzer, S. (2013). Hydrocarbon mobilization mechanisms from upper, middle, and lower bakken reservoir rocks exposed to CO. In *Paper presented at the SPE unconventional resources conference Canada, Calgary, Alberta, Canada, 2013/11/5.* https://doi.org/10.2118/167200-MS.

Honarpour, M. M., Nagarajan, N. R., Orangi, A., Arasteh, F., & Yao, Z. (2012). Characterization of critical fluid PVT, rock, and rock-fluid properties – impact on reservoir performance of liquid rich shales. In *Paper presented at the SPE annual technical conference and exhibition, San antonio, Texas, USA, 2012/1/1.* https://doi.org/10.2118/158042-MS.

Hughes, J. D. (October 28 2013). *Tight Oil: A Solution to U.S. Import Dependence?* Denver, Colorado, USA: Geological Society of America.

Hu, H. Y., Hao, F., Lin, J. F., Lu, Y. C., Ma, Y. Q., & Li, Q. (2017). Organic matter-hosted pore system in the Wufeng-Longmaxi (O(3)w-S(1)1) shale, Jiaoshiba area, Eastern Sichuan Basin, China. *International Journal of Coal Geology, 173,* 40–50. https://doi.org/10.1016/j.coal.2017.02.004.

Hu, H., Zhang, T., Wiggins-Camacho, J. D., Ellis, G. S., Lewan, M. D., & Zhang, X. (2015). Experimental investigation of changes in methane adsorption of bitumen-free Woodford Shale with thermal maturation induced by hydrous pyrolysis. *Marine and Petroleum Geology, 59,* 114–128. https://doi.org/10.1016/j.marpetgeo.2014.07.029.

Javadpour, F. (2009). Nanopores and apparent permeability of gas flow in mudrocks (shales and siltstone). *Journal of Canadian Petroleum Technology, 48*(08), 16–21. https://doi.org/10.2118/09-08-16-DA.

Javadpour, F., Fisher, D., & Unsworth, M. (2007). Nanoscale gas flow in shale gas sediments. *Journal of Canadian Petroleum Technology, 46*(10), 7. https://doi.org/10.2118/07-10-06.

Jiang, J., Shao, Y., & Younis, R. M. (2014). Development of a multi-continuum multi-component model for enhanced gas recovery and CO_2 storage in fractured shale gas reservoirs. In *Paper presented at the SPE improved oil recovery symposium, Tulsa, Oklahoma, USA, 2014/4/12.* https://doi.org/10.2118/169114-MS.

Kalantari-Dahaghi, A. (2010). Numerical simulation and modeling of enhanced gas recovery and CO_2 sequestration in shale gas reservoirs: A feasibility study. In *Paper presented at the SPE international conference on CO_2 capture, Storage, and utilization, New Orleans, Louisiana, USA, 2010/2011/1.* https://doi.org/10.2118/139701-MS.

Kalra, S., Tian, W., & Wu, X. (2018). A numerical simulation study of CO_2 injection for enhancing hydrocarbon recovery and sequestration in liquid-rich shales. *Petroleum Science*, 15(1), 103−115. https://doi.org/10.1007/s12182-017-0199-5.

Khatibi, S., Aghajanpour, A., Ostadhassan, M., Ghanbari, E., Amirian, E., & Mohammed, R. (2018). Evaluating the impact of mechanical properties of kerogen on hydraulic fracturing of organic rich formations. In *Paper presented at the SPE Canada unconventional resources conference, Calgary, Alberta, Canada, 2018/3/13*. https://doi.org/10.2118/189799-MS.

Khoshghadam, M., Khanal, A., Rabinejadganji, N., & John Lee, W. (2016). How to model and improve our understanding of liquid-rich shale reservoirs with complex organic/inorganic pore network. In *Paper presented at the SPE/AAPG/SEG unconventional resources technology conference, San antonio, Texas, USA, 2016/2018/1*. https://doi.org/10.15530/URTEC-2016-2459272.

Ko, L. T., Loucks, R. G., Ruppel, S. C., Zhang, T., & Peng, S. (2017). Origin and characterization of Eagle Ford pore networks in the South Texas upper cretaceous shelf. *AAPG Bulletin*, 101(3), 387−418. https://doi.org/10.1306/08051610635.

Kovscek, Robert, A., Tang, G.-Q., & Vega, B. (2008). Experimental investigation of oil recovery from siliceous shale by CO_2 injection. In *Paper presented at the SPE annual technical conference and exhibition, Denver, Colorado, USA, 2008/1/1*. https://doi.org/10.2118/115679-MS.

Kozeny, J. (1927). Über kapillare Leitung des Wassers im Boden. *Akad. Wiss. Wien*, 136, 271−306. https://doi.org/citeulike-article-id:4155258.

Ko, L. T., Zhang, T., Loucks, R. G., Ruppel, S. C., & Shao, D. (2016). Pore evolution in the Barnett, Eagle Ford (boquillas), and Woodford mudrocks based on gold-tube pyrolysis thermal maturation. In *AAPG search and discovery article*.

Kumar, V., Sondergeld, C. H., & Rai, C. S. (2012). Nano to macro mechanical characterization of shale. In *Paper presented at the SPE annual technical conference and exhibition, San Antonio, Texas, USA, 2012/1/1*. https://doi.org/10.2118/159804-MS.

Li, D., Xu, C., Wang, J. Y., & Lu, D. (2014). Effect of Knudsen diffusion and Langmuir adsorption on pressure transient response in tight- and shale-gas reservoirs. *Journal of Petroleum Science and Engineering*, 124, 146−154. https://doi.org/10.1016/j.petrol.2014.10.012.

Liu, F., Ellett, K., Xiao, Y., John, A., & Rupp. (2013). Assessing the feasibility of CO_2 storage in the New Albany Shale (Devonian−Mississippian) with potential enhanced gas recovery using reservoir simulation. *International Journal of Greenhouse Gas Control*, 17, 111−126. https://doi.org/10.1016/j.ijggc.2013.04.018.

Löhr, S. C., Baruch, E. T., Hall, P. A., & Kennedy, M. J. (2015). Is organic pore development in gas shales influenced by the primary porosity and structure of thermally immature organic matter? *Organic Geochemistry*, 87, 119−132. https://doi.org/10.1016/j.orggeochem.2015.07.010.

Loucks, R. G., Reed, R. M., Ruppel, S. C., & Hammes, U. (2012). Spectrum of pore types and networks in mudrocks and a descriptive classification for matrix-related mudrock pores. *AAPG Bulletin*, 96(5), 1071−1098. https://doi.org/10.1306/08171111061.

Loucks, R. G., Reed, R. M., Ruppel, S. C., & Jarvie, D. M. (2009). Morphology, genesis, and distribution of nanometer-scale pores in siliceous mudstones of the mississippian Barnett shale. *Journal of Sedimentary Research*, 79(12), 848−861. https://doi.org/10.2110/jsr.2009.092.

Lu, X.-C., Li, F.-C., & Watson, A. T. (1995). Adsorption measurements in Devonian shales. *Fuel*, 74(4), 599−603. https://doi.org/10.1016/0016-2361(95)98364-K.

Mastalerz, M., Schimmelmann, A., Drobniak, A., & Chen, Y. (2013). Porosity of devonian and mississippian New Albany shale across a maturation gradient: Insights from organic petrology, gas adsorption, and mercury intrusion. *AAPG Bulletin*, 97, 1621−1643.

Milliken, K. L., Rudnicki, M., Awwiller, D. N., & Zhang, T. W. (2013). Organic matter-hosted pore system, Marcellus formation (devonian), Pennsylvania. *AAPG Bulletin*, 97(2), 177−200. https://doi.org/10.1306/07231212048.

Modica, C. J., & Lapierre, S. G. (2012). Estimation of kerogen porosity in source rocks as a function of thermal transformation: Example from the mowry shale in the powder river basin of Wyoming estimation of kerogen porosity as a function of thermal transformation. *AAPG Bulletin*, 96(1), 87−108. https://doi.org/10.1306/04111110201.

Nelson, P. H., & Batzle, M. L. (2006). Petroleum engineering handbook. Vols (7−8). In *Single-Phase Permeability*. Richardson, TX: Society of Petroleum Engineers.

Pang, Y., Soliman, M. Y., Deng, H., & Xie, X. (2017). Experimental and analytical investigation of adsorption effects on shale gas transport in organic nanopores. *Fuel*, 199, 272−288. https://doi.org/10.1016/j.fuel.2017.02.072.

Pommer, M., & Milliken, K. (2015). Pore types and pore-size distributions across thermal maturity, eagle Ford formation, southern TexasPores across thermal maturity, Eagle Ford. *AAPG Bulletin*, 99(9), 1713−1744. https://doi.org/10.1306/03051514151.

Pu, H., & Li, Y. (2015). CO_2 EOR mechanisms in bakken shale oil reservoirs. In *Paper Presented at the Carbon Management Technology Conference, Sugar Land, Texas, 2015/11/17*. https://doi.org/10.7122/439769-MS.

Ross, D. J. K., & Bustin, R. M. (2007). Shale gas potential of the lower jurassic gordondale member, northeastern British Columbia, Canada. *Bulletin of Canadian Petroleum Geology*, 55(1), 51−75. https://doi.org/10.2113/gscpgbull.55.1.51.

Ross, D. J. K., & Bustin, R. M. (2009). The importance of shale composition and pore structure upon gas storage potential of shale gas reservoirs. *Marine and Petroleum Geology*, 26(6), 916−927. https://doi.org/10.1016/j.marpetgeo.2008.06.004.

Schepers, K. C., Nuttall, B. C., Oudinot, A. Y., & Gonzalez, R. J. (2009). Reservoir modeling and simulation of the devonian gas shale of eastern Kentucky for enhanced gas recovery and CO_2 storage. In *Paper presented at the SPE international conference on CO_2 capture, Storage, and utilization, San Diego, California, USA, 2009/1/1*. https://doi.org/10.2118/126620-MS.

Sheng, J. J. (2015a). Enhanced oil recovery in shale reservoirs by gas injection. *Journal of Natural Gas Science and Engineering, 22*, 252−259. https://doi.org/10.1016/j.jngse.2014.12.002.

Sheng, J. J. (2015b). Increase liquid oil production by huff-n-puff of produced gas in shale gas condensate reservoirs. *Journal of Unconventional Oil and Gas Resources, 11*, 19−26. https://doi.org/10.1016/j.juogr.2015.04.004.

Sheng, M., Li, G., Huang, Z., Tian, S., Shah, S., & Geng, L. (2015). Pore-scale modeling and analysis of surface diffusion effects on shale-gas flow in Kerogen pores. *Journal of Natural Gas Science and Engineering, 27*, 979−985. https://doi.org/10.1016/j.jngse.2015.09.033.

Shi, J.-Q., & Durucan, S. (2008). Modelling of mixed-gas adsorption and diffusion in Coalbed reservoirs. In *Paper presented at the SPE unconventional reservoirs conference, Keystone, Colorado, USA, 2008/1/1*. https://doi.org/10.2118/114197-MS.

Singh, H., & Javadpour, F. (2016). Langmuir slip-Langmuir sorption permeability model of shale. *Fuel, 164*, 28−37. https://doi.org/10.1016/j.fuel.2015.09.073.

Singh, H., Javadpour, F., Ettehadtavakkol, A., & Darabi, H. (2014). Nonempirical apparent permeability of shale. *SPE Reservoir Evaluation & Engineering, 17*(03), 414−424. https://doi.org/10.2118/170243-PA.

Sorensen, J. A., & Hamling, J. A. (2016). *Historical Bakken test data provide critical insights on EOR in tight oil plays*. The American Oil & Gas Reporter. https://www.aogr.com/magazine/cover-story/historical-bakken-test-data-provide-critical-insights-on-eor-in-tight-oil-p/.

Stalgorova, K., & Mattar, L. (2013). Analytical model for unconventional multifractured composite systems. *SPE Reservoir Evaluation & Engineering, 16*(03), 246−256. https://doi.org/10.2118/162516-PA.

Strapoc, D., Mastalerz, M., Schimmelmann, A., Drobniak, A., & Hasenmueller, N. R. (2010). Geochemical constraints on the origin and volume of gas in the New Albany shale (Devonian-Mississippian), Eastern Illinois basin. *American Association of Petroleum Geologists Bulletin, 94*(11), 1713−1740. https://doi.org/10.1306/06301009197.

Tabatabaie, S. H., Pooladi-Darvish, M., Mattar, L., & Tavallali, M. (2017). Analytical modeling of linear flow in pressure-sensitive formations. *SPE Reservoir Evaluation & Engineering, 20*(01), 215−227. https://doi.org/10.2118/181755-PA.

Tovar, F. D., Eide, O., Graue, A., & Schechter, D. S. (2014). Experimental investigation of enhanced recovery in unconventional liquid reservoirs using CO_2: A look ahead to the future of unconventional EOR. In *Paper presented at the SPE unconventional resources conference, The Woodlands, Texas, USA, 2014/4/1*. https://doi.org/10.2118/169022-MS.

Vega, B., O'Brien, W. J., & Kovscek, A. R. (2010). Experimental investigation of oil recovery from siliceous shale by miscible CO_2 injection. In *Paper presented at the SPE annual technical conference and exhibition, Florence, Italy, 2010/1/1*. https://doi.org/10.2118/135627-MS.

Vermylen, J. P. (2011). *Geomechanical studies of the Barnett shale, Texas, USA*. Doctor of Philosophy. The Department of Geophysics, Stanford University.

Wang, W., Shahvali, M., & Su, Y. (2015). A semi-analytical fractal model for production from tight oil reservoirs with hydraulically fractured horizontal wells. *Fuel, 158*, 612−618. https://doi.org/10.1016/j.fuel.2015.06.008.

Wan, T., & Liu, H.-X. (2018). Exploitation of fractured shale oil resources by cyclic CO_2 injection. *Petroleum Science, 15*(3), 552−563. https://doi.org/10.1007/s12182-018-0226-1.

Yuan, W., Pan, Z., Xiao, L., Yang, Y., Zhao, C., Connell, L. D., et al. (2014). Experimental study and modelling of methane adsorption and diffusion in shale. *Fuel, 117*, 509−519. https://doi.org/10.1016/j.fuel.2013.09.046.

Yu, W., Lashgari, H. R., Wu, K., & Sepehrnoori, K. (2015). CO_2 injection for enhanced oil recovery in Bakken tight oil reservoirs. *Fuel, 159*, 354−363. https://doi.org/10.1016/j.fuel.2015.06.092.

Zhang, T., Ellis, G. S., Ruppel, S. C., Milliken, K., & Yang, R. (2012). Effect of organic-matter type and thermal maturity on methane adsorption in shale-gas systems. *Organic Geochemistry, 47*, 120−131. https://doi.org/10.1016/j.orggeochem.2012.03.012.

Zhang, W., Xu, J., Jiang, R., Cui, Y., Qiao, J., Kang, C., et al. (2017). Employing a quad-porosity numerical model to analyze the productivity of shale gas reservoir. *Journal of Petroleum Science and Engineering, 157*, 1046−1055. https://doi.org/10.1016/j.petrol.2017.07.031.

Zhang, Y., Yu, W., Li, Z., & Sepehrnoori, K. (2018). Simulation study of factors affecting CO_2 Huff-n-Puff process in tight oil reservoirs. *Journal of Petroleum Science and Engineering, 163*, 264−269. https://doi.org/10.1016/j.petrol.2017.12.075.

Nomenclature

$a, b, c,$ and d Experimental coefficients (Equations 106–109)

a_{ii} Coefficients of quadratic terms

a_{ij} Coefficients of interaction terms

a_k Coefficients of linear terms

b Klinkenberg parameters (Equation 3.43)

b Slip coefficient (Equation 3.63)

$\mathbf{b}$ Body force per unit mass of fluid

c Compressibility

$c, c',$ and c'' Constant related to pore size distribution

c_b Bulk compressibility

c_f Pore volume compressibility

c_g Gas compressibility

c_t Total compressibility

d Grain diameter (Equation 3.2)

d Pore diameter (Equation 3.41)

d_m Normalized molecular size

d_p Local average pore diameter

h_m Thickness of matrix block

h_1 and h_2 Hydraulic heads

k Permeability

$\mathbf{k}$ Absolute permeability tensor

k_{app} Apparent permeability

k_D Darcy permeability or liquid permeability

k_{eff} Effective permeability

k_g Gas effective permeability

k_r Relative permeability

k_B Boltzmann constant ($1.3806488 \times 10^{-23}$ J/K)

l Thickness of sand (Equation 3.1)

l Original nondeformed dimension (Equation 3.88)

Δl Deformed dimension

n Number of normal fracture planes (Equation 2.5)

n Number of adsorption layers (Equation 3.35)

n Number of preferential hydraulic flow paths (Equation 3.66)

p_{avg} Mean pressure

p_c Critical pressure

Δp_c^* Relative critical pressure shift

p_{cb} Bulk critical pressure

p_{cog} Oil-gas capillary pressure

$\tilde{p}_{cog}$ Pseudo oil-gas capillary pressure

p_{cp} Pore critical pressure

p_{cwo} Water-oil capillary pressure

$\tilde{p}_{cwo}$ Pseudo water-oil capillary pressure

p_d Dew point pressure

p_L Langmuir pressure

p_o Saturation pressure of gas

p_{wf} Well flowing pressure

q_{sc} Rate at standard condition

r_p Pore throat radius

r_w Well radius

s Pore length

t Time

Δt Time step

$\mathbf{u}$ Displacement vector

v Superficial velocity

v_b Partial molal volume at boiling point

v_c Critical volume

w Weight fraction of clay

$x_D, y_D,$ and z_D Coordinates in reservoir model

x_k Input variables

y Mole fraction

y Response or objective function (Equations 4.28–4.29)

y' and z' Coordinate in organic matter

A Cross-sectional area

A_b Bulk surface area of porous media perpendicular to flow direction

B Formation volume factor

$\mathbf{B}$ Force per unit mass that accounts for gravity

C Concentration

C A constant related to net heat of adsorption (Equations 3.32 and 3.33)

$\mathbf{C}$ Tangential stiffness tensor

D Depth

D Diffusion coefficient

D_{ij} Binary diffusion coefficient between component i and j

D_{ij}^e Effective gas diffusivity

D_K Knudsen diffusion constant

E	Young's modulus or elastic modulus	α	Biot's constant (Equation 3.95)
E_L	Heat of adsorption for the second and higher layers	α	Parameter characteristics of system geometry or shape factor (Equation 2.5)
E_1	Heat of adsorption for the first layer	α	Tangential momentum accommodation coefficient (Equation 3.46)
F	Force		
$\mathbf{I}$	Identity matrix and parameter	α	Dimensionless rarefaction coefficient (Equation 3.63)
J	Total mass flux		
J_a	Advective mass flux due to pressure forces	$\alpha(\delta)$	Differential pore size distribution
		α', β', α'', and β''	Correction factors (Equations 3.9 and 3.10)
J_d	Molecular diffusion flux	β	Non-Darcy coefficient
J_D	Knudsen diffusive mass flux	β	Volumetric thermal expansion coefficient (Equation 3.101)
K	Bulk modulus		
K	Hydraulic conductivity (Equation 3.1)	$\overline{\beta}$	A dimensionless Langmuir adsorption constant
K_n	Knudsen number		
L_h	Hydraulic length or length of tortuous flow paths	β_{corr}	Non-Darcy correction factor
		β_{dry}	One phase non-Darcy coefficient
M	Molecular weight	β_r	Linear thermal expansion coefficient of solid rock
M	Axial modulus (Equations 3.36–3.39)		
M'	Solvent molecular weight	γ	Exponential of Euler's constant, 1.781 or $e^{0.5772}$ (Equation 2.7)
N	Moles of hydrocarbon per unit of grid block volume		
		γ	Interfacial tension (Equation 3.110)
N_{n_c+1}	Moles of water per unit of grid block volume	γ	Gradient of phase (Equation 4.4)
$N1_g$	Correlation parameter	γ	Permeability modulus (Equation 5.3)
Q_f	Flow rate at source or sink location		
R	Gas constant	δ	Collision diameter of gas molecule (Equation 3.42)
S_{gv}	Specific surface		
S_p	Paraffin saturation	δ	Pore diameter (Equation 3.12)
S_w	Water saturation	δ_R	Critical pore diameter
S_{wr}	Residual water saturation	ε	Lennard-Jones energy (Equation 3.81)
T	Temperature		
T	Matrix transpose (Equation 3.92)	ε	Strain (Equation 3.88)
T	Transmissibility (Equation 4.4)	ε_l	Strain at infinite pressure
T_c	Critical temperature	ε_x	Axial strain
ΔT_c^*	Relative critical temperature shift	ε_y	Transverse strain
T_{cb}	Bulk critical temperature	η	Diffusivity
T_{cp}	Pore critical temperature	θ	Tortuosity index of induced fracture network
V	Adsorbed gas volume (Equation 3.31)		
		θ_c	Contact angle
V	Grid block volume	λ	Interporosity flow coefficient
V_b	Bulk volume	λ	Mean free path
V_L	Langmuir volume	μ	Viscosity
V_L	Liquid molar volume (Equation 3.110)	ν	Poisson's ratio
		ρ	Fluid density
V_p	Pore volume	$\rho^0 D_{ij}^0$	Zero-pressure limit of density-diffusivity product
V_m	Maximum adsorption gas volume		
X	Dimensionless length	ρ_r	Reduced density
Z	Compressibility factor	σ	Stress

σ	Transfer coefficient or shape factor (Equation 4.9)	
$\boldsymbol{\sigma}$	Total stress tensor	
$\boldsymbol{\sigma}'$	Effective stress tensor	
σ_{ij}	Collision diameter	
σ_m	Mean total stress	
τ	Tortuosity	
τ_{gmf}	Matrix-fracture transfer in gas phase	
τ_{omf}	Matrix-fracture transfer in oil phase	
τ_{wmf}	Matrix-fracture transfer for water	
ϕ	Porosity	
ψ	Pseudopressure	
ω	Storativity ratio	
Ω_{ij}	Dimensionless collision integral of Lennard-Jones potential	

Subscripts

b	Bulk property
f	Fracture
g	Gas phase
i	Intrinsic property
i	Initial state
i, j	Component in mixture
k	Organic matrix (kerogen)
m	Matrix
o	Oil phase
p	Organic pores
w	Water phase
D	Dimensionless
F	Hydraulic fracture
U	USRV matrix

Index

Note: Page numbers followed by "f" indicate figures, "t" indicate tables.

Printed in the United States
By Bookmasters